Climatenomics

Climatenomics

Washington, Wall Street, and the Economic Battle to Save Our Planet

Bob Keefe

ROWMAN & LITTLEFIELD
Lanham • Boulder • New York • London

Published by Rowman & Littlefield
An imprint of The Rowman & Littlefield Publishing Group, Inc.
4501 Forbes Boulevard, Suite 200, Lanham, Maryland 20706
www.rowman.com

86-90 Paul Street, London EC2A 4NE

British Library Cataloguing in Publication Information Available

Library of Congress Cataloging-in-Publication Data

Names: Keefe, Bob, 1966- author.
Title: Climatenomics : Washington, Wall Street, and the economic battle to save our planet / Bob Keefe.
Description: Lanham : Rowman & Littlefield, [2022] | Includes bibliographical references. | Summary: "The battle against climate change is no longer just an environmental or social issue. As shareholders demand corporations protect assets against climate change and the economic impact of environmental disasters suck billions of dollars out of the economy, capitalism itself has become an ally. The economic impact of climate change is rattling the foundation of our economy at its very core. It's blowing up centuries-old industries from automobiles to oil and gas, creating new opportunities for investors and entrepreneurs. It's costing Americans billions of dollars each and every year. And most importantly, it's forcing politicians to pass long-overdue policies that will transform our businesses, our lives and our future like never before. The good news about this economic earthquake is that it just might be the thing that saves our planet. This is the first book to lay out how climate change has become an economic issue above all and how that has changed everything from the business to politics to the outlook for the future. Bob Keefe, executive director of E2, a national, nonpartisan organization dedicated to providing business perspectives on environmental issues, shows readers how this new reality will impact their industries, businesses, jobs, and communities and transform our country's economy. Climatenomics will be essential reading for anyone who cares about business, politics, or the future of our planet"—Provided by publisher.
Identifiers: LCCN 2022010187 (print) | LCCN 2022010188 (ebook) | ISBN 9781538168882 (paperback ; alk. paper) | ISBN 9781538168899 (epub)
Subjects: LCSH: Environmental policy—Economic aspects. | Climatic changes—Economic aspects. | Sustainable development.
Classification: LCC HC79.E5 K396 2022 (print) | LCC HC79.E5 (ebook) | DDC 333.7—dc23/eng/20220310
LC record available at https://lccn.loc.gov/2022010187
LC ebook record available at https://lccn.loc.gov/2022010188

∞™ The paper used in this publication meets the minimum requirements of American National Standard for Information Sciences—Permanence of Paper for Printed Library Materials, ANSI/NISO Z39.48-1992.

For Tammie, my pal and biggest fan,
and for Delaney, Grace, Carley, and
all who will inherit this place we will leave behind.

Contents

Foreword

By Tom Steyer

For the first 30 years of my career, I worked as a professional investor, starting my own investment firm, Farallon Capital Management, and growing it into one of the top firms in the US. In 2012, I left it all behind to dedicate my time, energy and resources to fighting the climate crisis. From the outside, this may have seemed like a drastic change and in many ways it was. But from my perspective I was doing what I had always done: listening to experts, evaluating future risks, and acting accordingly. The more I learned about the magnitude of the crisis and the effort it would take to remedy it, the more alarmed I grew at the lack of dialogue and awareness within American society about the state of the planet. I think about this crisis in terms of the old investing adage "strategy is easy, execution is hard." From where I stood in 2012, it seemed as though until Americans could recognize the urgency of the problem, even the strategy phase remained far off.

I quickly realized that our efforts to save the planet cannot be separate from the rest of society—we need to fundamentally restructure the world, starting with the economy. I found myself in good company as many share this perspective, including author Bob Keefe and the passionate business leaders who belong to E2, an organization I've known for a long time. The solution, we know, must be multi-faceted. A problem as huge as the climate crisis requires policy, business, and public participation. It will take Washington, Silicon Valley, Wall Street and Main Street to shift to a cleaner, more equitable and more resilient economy. If we do so, it will not only help address this crisis; it will create jobs, drive innovation and open new business and investing opportunities the likes of which we haven't seen since the dawn of the Internet age.

My first real foray into the politics of climate change came in 2010, when I co-led with George Schultz the campaign against Prop 23, an effort by out of state oil companies to kill California's landmark Global Warming Solutions

Act of 2006. Working on these measures, I saw firsthand the power of grassroots mobilization in educating voters and effecting change. So, in 2012, I founded NextGen Climate, now NextGen America, to educate and mobilize a new American electorate that would push for bold climate legislation. In 2015, NextGen ran a campaign urging Democratic candidates to embrace the goal of powering America with more than 50 percent clean and carbon-free energy by 2030. At the time, this goal was described by many as too bold and going too far. But public support for investing in clean energy has rapidly increased and just six years later, the Biden Administration embraced a target of 100 percent clean energy by 2035.

Most of us understand the need for climate action and a cleaner economy. As of this writing, three-quarters of all voters, including 92 percent of Democrats, 75 percent of Independents, and 57 percent of Republicans, agree the U.S. should work with other countries to combat climate change and reduce global greenhouse gas emissions. And over two-thirds of voters, including nearly all Democrats (90 percent), a majority of Independents (62 percent), and a plurality of Republicans (47 percent), agree that the U.S. should lead the world in addressing climate change so other countries will follow suit. At the same time, corporations, entrepreneurs and investors have come to similar conclusions, and the costs of clean energy have plummeted. We have completed the strategy portion. Now, with the private sector at the helm, we need to move to execution.

We cannot wait any longer. If we continue to emit at our current rate, our carbon budget to limit global warming to 1.5C will expire in under eight years. We need to decarbonize our economy, electrify everything, strengthen our grid, and begin taking carbon out of the atmosphere. And we need to make sure we do this in a just and equitable way. This won't be easy—the latest IEA report estimates we need a global investment in the energy transition of $4 trillion per year for the next 30 years.[1] Currently, we are investing just under $1 trillion, leaving a $3 trillion per year investment gap.

We shouldn't wait for the public sector to solve this problem—we know we need major capital and serious creativity from the private sector to succeed. This moment demands creativity, not business as usual. Beating this crisis requires innovation, entrepreneurship and disruption. It will take everything from the garage startup with an innovative new idea to the international conglomerate rebuilding their business model to solve this crisis.

In 2021, I sat down with my old friend and business partner Katie Hall to brainstorm how we could meaningfully contribute to accelerating climate solutions. After surveying the landscape, everything from traditional advocacy to politics to the private sector, we determined that whatever we built had to straddle all three worlds. Ultimately, we determined that we could meaningfully participate in meeting this challenge by helping to mobilize and deploy the capital and expertise necessary to accelerate and scale equitable climate solutions.

And so I returned to my roots as an investor and alongside Katie co-founded Galvanize Climate Solutions, a mission driven investment platform working to invest in meaningful solutions and mobilize the private sector at large to work alongside state, federal and international governments to solve this crisis. We aim to prove that investing in the clean energy ecosystem works for the planet, and for its investors as well. We believe that as the need for clean energy investments becomes increasingly obvious, the entire finance sector and business world will join the effort to accelerate the transition to clean energy. No one company can beat this crisis alone; this moment calls for a fundamental restructuring of the role of the private sector.

But the private sector cannot and will not do this alone. To accomplish broad societal change, we need the participation, imagination and determination of government, civil society and the private sector working together. Meeting this moment means prioritizing radical collaboration and cooperation between companies and industries. While rebuilding the world presents a historic investment opportunity, waiting for each sector to develop sustainable alternatives without collaborating first wastes time and resources. We need public-private partnerships that rival the partnerships we saw around World War II and unprecedented collaboration between the industries we need to transform.

When I talk about this crisis, people frequently ask me some variation of "I know the crisis has arrived—what can I personally do?" The obvious answer for everyone is simply to vote for climate champions. Solutions to this crisis won't materialize without policymakers and businesspeople working together, so voting and participating in our democracy have to be a priority for every citizen. But elections don't happen every day, so what else?

The biggest opportunity lies with private sector actors. To any investors or asset managers reading this book, I'd encourage you to push your leadership on their climate plans. You can read examples of businesses that are leading on the pages that follow. The future belongs to clean energy. Investing in protecting the planet does not call for just concessionary capitalism or philanthropy. Rebuilding the world presents a massive commercial opportunity, arguably the largest investment opportunity of our time. The investors and institutions that hold onto the fossil fuel economy will lose out on the prosperity that the clean energy transition will inevitably bring and jeopardize it for everyone else.

The innovative spirit and entrepreneurial drive that allowed the American economy to flourish over the last century must be redirected to the defining crisis of our generation. Those that embrace this new way of thinking—what I call Movement Capitalism—will succeed financially and help mitigate major environmental damage at the same time. Companies no longer have to choose between adhering to their fiduciary responsibilities and saving the world.

A decade after I decided to dedicate my time, energy and resources to fighting this crisis, I can see more clearly than ever that our climate and economy are not only entangled but deeply and increasingly reliant upon one another. It's *climatenomics*. And to win this fight, we must all work together—businesses, policymakers and the public. We still have a viable path to a safe and healthy future but we must work quickly. We don't have any time to waste.

Tom Steyer, a 2020 Democratic presidential candidate and co-chair of California Gov. Gavin Newsom's Business and Jobs Recovery Task Force, is the co-executive chair of Galvanize Climate Solutions, a mission driven investment platform that will provide the capital, expertise and partnerships necessary to produce and scale vital and urgent climate solutions.

Preface

By Manish Bapna

Before I was an environmentalist, I was an economist.

Seven years as an economist at the World Bank, another four as executive director of the Bank Information Center and a few years as a strategy consultant taught me the importance of business, policy and finance in solving the world's biggest problems.

Today, the world is facing an existential problem: climate change. To solve it, it, we're going to need the help of business, policy, finance–and much more.

Climate change has always been an economic issue. Just ask any farmer who watched their crops and income wither under a punishing drought; any business owner who lost their life's work in a matter of minutes to hurricane winds or torrential flooding; any real estate investor or homeowner whose assets went up in smoke amid the flames of a hellish wildfire. But also listen to any of the nation's more than 3 million clean energy workers who today make a good living in steady jobs that didn't even exist a couple of decades ago. Talk to leading lawmakers and officials who understand how the power of policy can shape a cleaner economy, create jobs, drive economic growth and make our country safer, more inclusive and more competitive. Listen to the investors and entrepreneurs who through innovation are revolutionizing the way we produce our electricity, grow our food and travel from point A to point B.

As executive director of E2 (Environmental Entrepreneurs), and before that as an award-winning business, technology and political journalist, that's what Bob Keefe did. With a journalist's eye, an environmentalist's passion and a storyteller's skill, Bob lays out in *Climatenomics* the economic costs of climate change, the economic benefits of climate action and how policy and business are poised to rebuild our economy and help save our planet.

The economics of climate change are increasingly evident. Not long ago, business leaders and policymakers alike thought the choice was black or white: We could either focus on creating jobs and driving economic growth, or we could focus on climate action and protecting the planet.

Today, we know that's a false choice.

In 2014, I was privileged to be a part of a team that set up the Global Commission on the Economy and Climate—a group of former heads of state, ministers, CEOs and other influential leaders—that was tasked to examine how countries around the world can advance robust economic growth while also tackling climate change. The New Climate Economy flagship report endorsed by the Commission was about the size of this book, but its ultimate conclusion was as simple as it was clear: The economic benefits of climate action are greater than ever, while the economic costs of inaction are growing increasingly worse. If the world shifted decisively and immediately toward a low-carbon economy, the Commission found that it would unlock $26 trillion in global economic benefits. Shifting to clean energy and pursuing other interventions to reduce greenhouse gas emissions while also investing in climate adaptation would create more than 65 million jobs, avoid more than 700,000 premature deaths from air pollution and raise $2.8 trillion in revenues worldwide. Doing all of this with equity and inclusiveness at the center could also create a just economic transition.

Today, the benefits are even bigger, easier to unlock, and needed more than ever.

This is true in the United States, which is seeking a better growth story. We can no longer afford to pin our future economic growth to the fossil-fueled, high-carbon model of the past. In 2021, the costs from climate-related disasters from wildfires, hurricanes, floods and droughts was more than $145 billion, according to the National Oceanic and Atmospheric Administration (NOAA) – nearly 50 percent more than the previous year alone. Our economy wasn't built for economic hits like that. We need to shift from a low-growth, high-carbon economy to a high-growth, low-carbon economy.

We also cannot continue to allow the inequitable economic model that defined our past to persist, where only the rich can afford electric vehicles, solar panels and the benefits of energy efficiency, while low-income and communities of color unjustly bear the biggest burden of pollution and climate-related disasters. We simply can't deploy enough clean energy and reduce carbon emissions quickly enough if only the wealthiest and most well-off in our economy are part of the transition and others are left behind.

The good news is that we can create a better growth story–one that includes millions of new jobs in clean energy and clean transportation; trillions of dollars in new investments in clean energy and climate-tech businesses and innovation and a transformation of our economy for the better like we haven't seen in generations. To unlock that growth, to create that better story, three things are necessary. We need ***policies*** to frame our goals, set standards and provide initial impetus and investments that send the market signals necessary for the private sector to move. We need ***businesses*** to act by building and delivering the products we need–whether it's more affordable electric vehicles or climate-tech solutions for agriculture, buildings, heavy industry and power generation – while also sustainability leading the way in the transition to a cleaner economy. And we need ***finance*** – investors, banks, other funders–to provide the capital to businesses and consumers who are leading this transition to a cleaner economy, while immediately ending financial support of companies and industries that seek to keep us tethered to our more costly, unsustainable carbon-based past.

In ways like few others can, Bob Keefe in *Climatenomics* explains the importance of these interconnected elements of policy, business and finance to creating a new growth story for the United States. His experience as a journalist, as an indefatigable advocate for our environment and the leader of a national organization of more than 11,000 business leaders who understand the clear connection between economy and environment gives him rare insight into Washington, Wall Street and the economic battle to save our planet.

As *Climatenomics* lays out, the Biden administration's climate and clean energy agenda will take the United States a big step forward–as long as we don't let partisan politics and short-sighted ideology get in the way. International climate agreements forged in Paris and reaffirmed in Glasgow will help get us further, even though much more needs to be done. Meanwhile, the continued leadership of states and cities remain critical to filling in the gaps and keeping momentum moving as the wheels of federal and international policy grind slowly forward.

Businesses also have a pivotal role to play. These include major corporations ranging from Google to General Motors that can impact change both through their own procurement and supply chain practices and by bringing to us the cutting-edge innovations a cleaner, better economy requires. But they also include entrepreneurs and mom-and-pop businesses across America, like those you'll meet in the pages that follow.

In finance, we are seeing significant shifts, but much more needs to be done. Big banks and investment funds can lead the way – but they also must be held accountable for the promises they make. Meantime, the importance of early-stage funding, whether it comes from upstart Silicon Valley venture capital firms or the well-heeled wizards of Wall Street, will also be critical.

The hard truth is that we need all of this and more. Climate math is unforgiving. We know we need to limit global warming to no more than 1.5C by the end of the century to avoid the most catastrophic impacts of climate change, but as I write this, we're on a trajectory of 2.4C of warming, and this assumes that promises made are promises kept. We must focus on implementation and ambition; we need to implement climate commitments already made while at the same time making more ambitious commitments. To succeed, we must deploy policy, business and finance to shift quickly and responsibly to a cleaner, stronger and more inclusive low-carbon economy. And always, we must be vigilant and push back against politicians and business interests who would risk the future of our country and our planet for short-term profits and political points.

This is in the best interests of all of us, of course–and that includes business and industry. Each year, the World Economic Forum releases an annual survey of business, finance, policy and world leaders. A decade or so ago, climate change issues hardly registered. In WEF's 2021 Global Risks Report, respondents ranked "climate action failure" as the second most pressing risk facing the world – higher in terms of risk and impact than weapons of mass destruction, terrorism or other economic issues. Only "infectious diseases" ranked higher in the annual report–understandably so as the world was still mired in the COVID-19 global pandemic.

Despite setbacks, leadership in Washington and in state capitals by and large now understands the critical need for impactful policy to address climate change. Thanks to business and innovation, technology is catching up with our needs, making clean energy the cheapest power available in most parts of the country and creating cleaner, more affordable solutions for our homes, on our roads and down on our farms. Now it's time for speed and for scale, which investors and businesses can deliver. We can't wait any longer.

Climatenomics chronicles the economic costs of climate inaction and identifies the economic benefits of climate action at this critical time of transition in our country and across the planet. It is a must-read for businesspeople, policymakers, politicians and anyone who cares about our environment or our economy, and a manual for a cleaner, brighter, better future for all of us.

Manish Bapna is president and chief executive officer of the Natural Resources Defense Council (NRDC). He previously served as executive vice president and managing director of the World Resources Institute; executive director of the Bank Information Center and senior economist for the World Bank.

Acknowledgments

I am blessed beyond measure.

For most of my life, I've somehow managed to make a living meeting interesting people, learning interesting things, telling interesting stories, and hopefully doing a little public good. As an old journalism colleague of mine used to say, who wouldn't want a job like that?

We can never properly thank all the people who help us along our way. But for this particular project, I can thank a few, and I can humbly apologize to others I may have accidently forgotten.

I begin with the incredible staff of E2, the organization I've been fortunate to be a part of for more than a decade and lucky enough to lead since 2015. I continue to be amazed at how big of an impact our small but mighty team makes in Washington and in statehouses across the country. Thanks to Crew E2's hard work, we have made business and the economy an integral part of just about any discussion about climate and environmental policy. We've changed the national narrative around clean-energy jobs. We've connected countless businesspeople and lawmakers, advanced critical laws and legislation, and together have done and continue to do good for our economy and our environment. Even better, we usually find a way to have fun doing it. Special thanks to Sandra Purohit for helping keep me straight on policy, Michael Timberlake for his help with charts and graphs and numbers, and Gail Parson for her partnership and her steady hand steering the ship whenever I needed to step away from the tiller for this endeavor.

Thanks also to Jeff Benzak, for his help with editing and for being alongside me more than a decade ago when we decided that clean jobs count, and that when our nation's leaders think about climate and clean energy, they should think about jobs. Thanks too to Phil Jordan and his team at BW Research Partnership. A trusted partner and friend, Phil has a heart of gold and knows more about the growth of the clean-energy workforce in America than anyone.

Nothing good can be said about E2 without mentioning Nicole Lederer, its chair and co-founder. I am forever grateful to Nicole for so much: For always supporting me and my ideas—including the idea for this book; for her vision, passion, and tenacity and for her guidance and friendship. Without Nicole and Larry Orr, there would be no E2 today, and without E2, I never would have been in the position to do what I do, including writing this book. Thank you both. Likewise, E2 could do nothing without the support of its members and donors. I'm thankful for each and every one of you, especially those mentioned in the pages that follow for providing your thoughts, insight, and time. If you're a businessperson who cares about the environment and you don't belong to E2, you really should.

At NRDC, I owe thanks to Rob Perks for his support and Ed Yoon for his leadership; to Ed Chen, always, for giving me the opportunity to begin with to keep telling good stories at the intersection of economy and environment, and to Jenny Powers, Lisa Goffredi and others for their camaraderie and support from the very beginning. Special thanks to NRDC President and CEO Manish Bapna, who understands the connection between economy and environment better than most, for his help, support, and leadership. Thanks too to NRDC's previous president, National Climate Advisor Gina McCarthy, who has given all of us a stronger voice in the fight against climate change and hope for the future. Tom Steyer was gracious to write the forward for this book, for which I'm very grateful. I'm even more grateful for his vision and passion as a business leader who truly understands both the economic and environmental importance of protecting our planet.

I am forever grateful and indebted to my longtime colleague and friend Bob Deans. A master communicator, gifted writer, and Southern gentleman to his James River–soaked core, I could thank Deans for a million things, but I know he'd wisely counsel me to limit it to just three because neither of us would likely remember any more than that, at least not at our age. Thank you, Deans, for showing me the path from journalism to environmentalism; for connecting me with Jon Sisk and the talented team at Rowman & Littlefield; and for your support, guidance, and friendship over all these years. That's sort of three.

To Tammie, thanks for always believing in me and for supporting my lifelong bad habit of spending countless hours stringing words together in my unsatiable desire to try and tell a good story. I love you always.

Lastly, we should all be grateful to those in government, business, advocacy, and the media who are on the front lines of the fight against climate change, and to all who work every day to make our world a little bit better.

Carlsbad, California
December 2021

Chapter One

(Climate) Change Happens

May 26, 2021, was a bad day to be in the oil and gas business.

At ExxonMobil's annual meeting, shareholders demanding that the company move away from fossil fuels and expand into clean energy shook up the company's board of directors by electing three of their own to new board seats. Over at Chevron's annual meeting, which was occurring at nearly the exact same time, a shareholder resolution to force the 142-year-old petroleum company to cut its carbon emissions by shifting to cleaner energy overwhelmingly passed with more than 60 percent of shareholders voting in favor of it. Meanwhile, a court in The Hague ruled in favor of environmentalists and others that Royal Dutch Shell was on the verge of breaching its corporate obligations to reduce carbon emissions. In a first case of its kind, the court ordered the oil giant to reduce emissions by 45 percent by 2030 compared to 2019 levels.

On its surface, the stunning same-day trifecta for climate activists would seem to be just the latest sign of the growing pressure by environmentalists to force companies into action on climate change. And that's how many media outlets characterized it.[1]

But the truth is, the developments were as much about economics as they were about climate activism. Shareholders backing the ExxonMobil revolt weren't just well-to-do tree huggers or young millennials making a stand for their future. They also included Vanguard, one of the world's largest mutual fund companies with $7 trillion in assets built on the pensions and retirement accounts of American workers; State Street, the stoic financial firm that operates the second-oldest bank in the country, and BlackRock, the world's biggest asset manager. At Chevron, unhappy shareholders cited the need to "protect our assets against devastating climate change," pointing out in their proxy proposal that climate risks "are a source of financial risks."[2] The Dutch court in its ruling against Shell noted prominently that climate change will "impact the ecology, but also the economy," citing specifically the loss of biodiversity and the impacts on two cornerstones of the Netherlands' economy, fisheries and agriculture.[3]

The oil companies learned that day that something had changed. They learned what the rest of the world is learning: That climate change is no longer just an *Inconvenient Truth,* like we discovered in the 2000s. It's no longer just a social issue to be protested in the streets like activists around the world did in the 2010s. In the 2020s, climate change is an economic issue. And that changes everything.

The economics of climate change—*climatenomics*—is rattling the foundation of our economy to its very core. It's disrupting centuries-old industries from petroleum to automobiles to utilities and finance. It's creating new opportunities for investors and entrepreneurs. And it's sucking money out of the pockets of every American with every wildfire, hurricane, flood, and drought made worse by climate change. Perhaps most importantly, now that money and jobs are involved, climatenomics is finally forcing politicians to pass policies that will forever transform our businesses, our lives, and our future.

The seismic shake-up of the petroleum industry in the spring of 2021 is just one indication of how the economics of climate change is forcing change in businesses, governments, and ultimately the world in ways where climate activism, environmental documentaries, and political lobbying fell short.

Fact is, for all the problems that come with burning oil and gas and the deserved scrutiny of the industry, the Big Oil companies would have never changed if not for climatenomics. Shareholders don't revolt when stock prices are strong. Management doesn't get called on the carpet when they have forward-looking business plans that show a path toward long-term stability and success. Just ask the management of Exxon. For decades, the company rode the world's insatiable demand for oil and gas to the highest pinnacle of the petroleum industry's energy monopoly. By the early 2000s, Exxon was the most valuable company in the world, gushing record profits year after year. To protect itself and its profits, it used its seemingly limitless cash and power to fund climate disinformation campaigns[4] and support politicians who would later place its executives in the highest positions in government (remember Exxon CEO Rex Tillerson, U.S. secretary of state under President Trump?) and squash lawsuits by environmental and citizen groups. But when the world began moving toward cleaner energy, Exxon remained stubbornly stuck in the past. It doubled down on oil and gas while downplaying the financial risks of climate change and the threat to its core business that comes with every electric vehicle rolling off the assembly line and every solar and wind installation popping up on the horizon. And then the COVID-19 pandemic stopped the world in its tracks, sending the demand for Exxon's products plummeting. The company—and its shareholders—paid the price. By the time of its annual meeting in May 2021, Exxon's shares were worth nearly half as much as they were worth just five years earlier. The company lost more than $22 billion in 2020. Its market capitalization was half what it was in 2008.

For years, investors had appealed to Exxon to get its corporate head out of the sand and address the financial risks of getting left behind in a world shifting from fossil fuels to cleaner energy. Representatives of Vanguard, for instance, met with company executives and board members more than a dozen times during 2020, including hurried last-ditch meetings before, during, and even after the May 2021 proxy vote.[5] Exxon board members didn't listen. Ultimately, seeing little opportunity to change the company otherwise, and feeling the pressure from its own clients, Vanguard joined a small hedge fund in its proposal to elect a new slate of board members who would listen to its shareholders and begin to make changes to move the company away from the dying business of oil and gas and toward the growing business of cleaner energy.

"Vanguard expects portfolio companies and their boards to be competent about material risks," the mutual fund firm wrote in a note to its clients.[6] "With climate change, this includes an expectation of appropriate risk oversight, mitigation strategies and practices, and effective disclosure to the market of how the board oversees climate-related strategy and risk management." Exxon's current board wasn't doing its job, Vanguard said in voting its 344 million shares in favor of a new slate of directors.

Among major investment firms, few have been more vocal about the financial risks of climate change than BlackRock. Like their peers at Vanguard, BlackRock managers also met with Exxon's board and top executives repeatedly to try to get the company to face up to the economic costs and risks associated with its core business and do more to position itself to reap the growing economic rewards that were becoming more apparent every year with clean-energy's growth. In particular, BlackRock was concerned Exxon was investing much less than its competitors in research, development, and deployment of lower-emissions technology. For BlackRock, Exxon's failure to invest in the future while continuing to pump money into yesterday's technology meant its business—and its stock price—would continue to decline. New leadership at the board level was needed.

"In our view, Exxon and its Board need to further assess the company's strategy and board expertise against the possibility that demand for fossil fuels may decline rapidly in the coming decades," BlackRock wrote to its clients.[7] "The company's current reluctance to do so presents a corporate governance issue that has the potential to undermine the company's long-term financial sustainability."

Leading the unprecedented investor uprising against Exxon was Chris James, a serious, fifty-one-year-old investment manager and founding partner of the little hedge fund Engine No. 1. James isn't your typical environmental activist shareholder.[8] He grew up in southern Illinois in what was once the center of the Midwest's coal industry. Then he attended Tulane University in Louisiana, giving him a front-row seat to the growth, potential, and importance of offshore oil. He owned and operated a coal mine in Illinois and an oil and gas storage tank company before moving to Silicon Valley and making a fortune investing in biotech and medical technology companies. Like so many other business people, James eventually realized that he could make money and also make a difference in the world. He began to realize that his personal values and the future he wanted to leave to his children didn't align with his investment portfolio—or his financial goals as an investor. He decided to use his hedge fund, whose name is a double-entendre nod to the fire station near his firm's San Francisco office and to the urgent threat of global warming, to pressure companies to change. After long discussions with partners and fellow investors, he made the decision to begin with the biggest and most problematic company of them all.

In launching his crusade against Exxon, James didn't talk about climate change or the environment. He talked about something he knew about most: Making money. And for Exxon to make more money, according to James, it needed to reduce its investments in oil and gas and develop plans to invest in and profit from the boom in clean energy. Applying the lessons learned from his own history investing in oil, gas, and technology, James contended that the company also needed to come to grips with the fact that its core business was going the way of whale oil and steam engines and that if it wanted to survive and prosper in the long term, it needed to adapt and embrace technological change.

"If we're right on getting Exxon to mitigate these impacts, the stock should go up," James told the *Wall Street Journal*, "and maybe Exxon does have a future."[9]

In picking nominees to fill the new seats at Exxon, James and team were careful. They spent months meeting with and vetting potential board nominees, rejecting climate firebrands and anyone who could've been characterized as caring more about polar bears than profits. In the end, they landed on four candidates, three of whom had long histories in the oil and gas business and track records of profit-making that were front-and-center in Engine No. 1's quixotic quest to shake up Exxon's board.[10] Nominee Kasia Hietala got her first job as a crude oil sourcing manager for Finnish oil and gas giant Neste Oyj, working her way up the ranks to become the company's executive vice president of renewable fuels, where she oversaw a quadrupling of profits during her five years at the post. Gregory Goff was an old-school petroleum business legend, with a forty-year career that included stints as executive vice chairman of Marathon Petroleum Corp. and president and CEO of Andeavor, formerly known as Tesoro Corp., one of the world's biggest petroleum refining companies. During his eight years running Tesoro, Goff produced profits that were the envy of the petroleum business, generating shareholder returns of more than 1,200 percent, compared with 55 percent for the rest of the industry. The company's stock price soared from about $12 per share to more than $150 per share under Goff's management. If Exxon shareholders wanted more profits and higher stock prices, there was nobody like Goff who could help it deliver. And then there was board nominee Andy Karsner, who was raised in Texas and worked as an analyst for a Houston oil, gas, and power plant company before being tapped by fellow Texan President George W. Bush to become assistant secretary of the Department of Energy. Outspoken and unabashed, Karsner had figured out over a career in politics, technology, and energy how to masterfully walk the sharp edge between fossil fuels and renewable energy. When Bush, himself a former oil man, proclaimed as president in 2006 that "America is addicted to oil," he turned to Karsner at the Department of Energy to help kick the habit. Karsner helped launch DOE's Advanced Research Projects Agency-Energy (APRA-E) and through it helped advance solar and wind innovation, but also research into fracking, algae-based fuels, and carbon capture and storage, appeasing both renewable energy proponents and oil and gas executives and investors. Along with their deep connections to the oil industry, however, all three of the candidates had also come to realize the economic and environmental consequences of climate change—and the role of the petroleum industry in it. After leaving Neste, Hietala started a consulting business to help big companies become more sustainable. Goff remained an oil industry booster as he rode off into retirement at his ranch in the Texas Hill Country, but he also recognized that to keep his beloved industry growing, it had to face up to the realities of a changing energy business and consider new, cleaner ways of producing its products. Karsner after leaving government went on to invest in wind and technology companies and later joined the innovation arm of Google parent Alphabet Inc. as its self-proclaimed "space cowboy."

In an interview with the *Washington Post* after his election to Exxon's board, Karsner explained his perspective, one that seemed to mirror that of his fellow new board members as well. "I don't go into this looking for villains," he said.[11] "I'm looking for constructive, candid engagement as a way to make this a better, more profitable, more sustainable company."

Eight months after its board shake-up, Exxon Mobil in January 2022 announced it had set a goal to reach net-zero carbon emissions from its operations by 2050. The United States' biggest petroleum company stopped notably short by leaving out of its goals so-called "Scope 3" emissions from the burning of its products. Still, it did pledge that its "ambition" was to zero-out emissions produced by its own operations and from the energy it buys to power those operations. Exxon's monumental announcement would not have happened when it did if not for the shareholder uprising led by Engine No. 1. Chevron and Shell also made similar announcements months earlier.

The fact that climate change is now a dollars-and-cents business issue isn't just upending centuries-old industries and companies like Exxon. It's upending our entire economy, cutting into America's gross domestic product each year, costing taxpayers money with every wildfire, hurricane, flood, and drought, setting the stage for even greater economic losses in the future.

Between 2017 and 2021, climate-related weather disasters caused more than $742 billion in damage to the U.S. economy, including a staggering $145 billion in 2021 alone, according to the National Oceanic and Atmospheric Administration. Spread across the calendar, climate-related disasters sucked about $400 million out of our economy on a daily basis in 2021—more than the U.S. spent daily on the war in Afghanistan. The losses of 2021 were on top of $100 billion in damages from weather and climate-related disasters in the previous year, ushering in a new decade in which the stark reality of the costs of climate change hit home in no uncertain terms in the United States.[12]

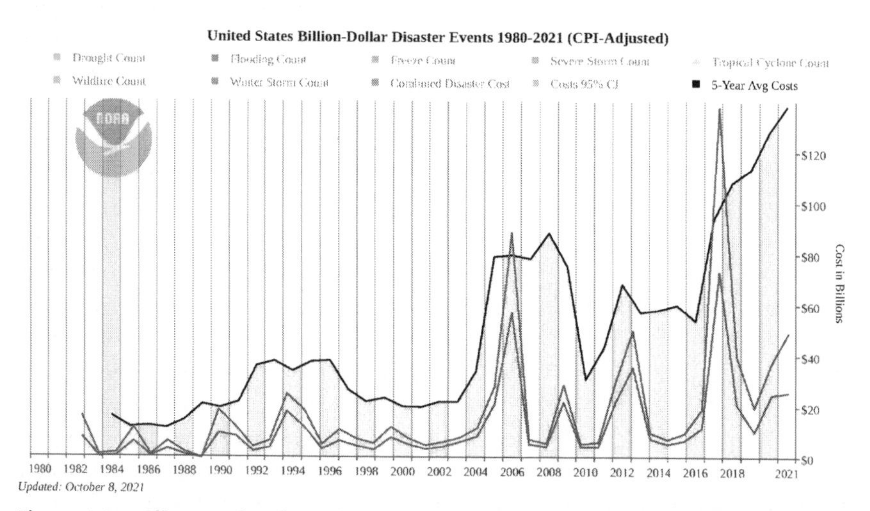

Figure 1.1. Climate-related weather disasters totaled $145 billion in 2021, up nearly 50 percent from the year before. Between 2017 and 2021, climate-related disasters sucked more than $742 billion out of the U.S. economy. *Source*: NOAA National Centers for Environmental Information (NCEI), U.S. Billion-Dollar Weather and Climate Disasters (2021).

Nationwide, nearly fourteen million acres and eighteen thousand homes and other structures—mainly in the West—went up in smoke in 2020, and the government spent a record $3.6 billion on suppression costs, according to the National Interagency Fire Center.[13] In Oregon, four thousand homes were destroyed, leaving residents of the Rogue Valley and other parts of the state living in borrowed tents in homeless camps.[14] In Colorado, three of the largest fires in state history occurred in 2020, including the aptly named East Troublesome Fire that burned through ten thousand acres of dead trees and grasslands east of Denver near Rocky Mountain National Park. In an indication of the bizarreness of climate change, firefighters battled that particular inferno during the middle of a major snowstorm. It was only with the help of snow, ice, and below-freezing temperatures that they finally contained the fire—a month and a half after it started.[15] In California, a freakish "lightning siege" lasted for four days in August 2020, zapping the state with more than twelve thousand lightning bolts in a single week, including as many as two hundred in thirty minutes' time that lit up the Bay Area like a giant electrostatic lamp.[16] The lighting sparked hundreds of wildfires—some of the nearly ten thousand wildfires that plagued California that year alone. Among them was a conflagration in Lassen County that burned so high and hot that it created its own weather system, prompting the National Weather Service to issue its first-ever "fire tornado warning"[17] as terrified residents faced down swirling furies of flames and wind like something unleashed by a supernatural villain in a Marvel Comics story. A few weeks later, President Trump visited the area, where he suggested that climate scientists didn't know what they were talking about and claimed the answer to wildfires was as simple as raking forest floors. For beleaguered California Gov. Gavin Newsom, it was a breaking point. "I have no patience for climate-change deniers," Newsom said angrily in September 2020. "It's inconsistent with the reality on the ground, the facts."[18]

By the end of the year, the facts on the ground were clear, devastating, and expensive. Across California, Oregon, Washington, and Colorado, fourteen thousand buildings were destroyed and nearly fifty people were killed. The $16 billion in direct economic losses from Western fires[19] that year were staggering enough, but that only included the direct capital losses for buildings, homes, and property. Based on previous research, the total economic costs are much, much higher. In 2018, for instance, California wildfires caused about $28 billion in direct capital impact (burned homes and buildings), but when other costs such as supply chain interruptions, lost productivity, and health impacts from air pollution were considered, the truer cost was more than five times that amount, researchers at the University of California, Irvine and elsewhere found. In 2018, they determined, the truer economic cost of wildfires in California alone was nearly $150 billion.[20]

While the West burned, other parts of the country were soaked and pummeled by storms. In 2020, the busiest Atlantic hurricane season on record brought with it thirty named storms, including thirteen hurricanes.[21] Among them was Isaias, which on paper went down as a relatively mild Category 1 storm in August 2020 but went on to spawn nearly forty devastating tornadoes from rural Bertie County, North Carolina, to downtown Philadelphia as it churned up the Atlantic Seaboard. One tornado clocked in with winds of 145 miles per hour, making it one of the strongest hurricane-generated tornadoes ever recorded in the United States.[22] A month later, Hurricane Laura made landfall in Louisiana with sustained winds reaching 150 miles per hour, tying it with the strongest hurricane ever recorded in Louisiana history. By the time Laura and its Category 4 winds were done whipping Louisiana and surrounding states, it had caused more than $13 billion in damage—about thirteen times as much money as Mardi Gras brings into the state annually. Severe storms weren't just isolated to the coasts in 2020, however. In Nashville, right as the "Super Tuesday" elections were about to begin in March, a massive "supercell" generated fifteen tornadoes that whipped grounded airplanes and helicopters around like little toys at John C. Tune Airport, blasted the belltower off the 113-year-old East End United Methodist Church, destroyed hundreds of homes, killed twenty-five people, and caused $1.6 billion in damage.[23] And in Iowa, a hurricane-force derecho, the likes of which had never been seen before, brought with it a line of tornadoes that in places stretched seventy-five miles across, tearing up homes, office buildings, and ten million acres of crops, causing an estimated $7.5 billion in damage and making it the costliest thunderstorm-related disaster in U.S. history. "The destruction," said Iowa Gov. Kim Reynolds, "was indescribable."[24]

She could've been speaking for the entire country in 2020.

And then, in 2021, it got worse.

Extreme drought and record temperatures combined to add even more kindling to the tinderbox that was the Western half of the United States. At one point in the summer of 2021, fully 90 percent of the West was in drought. Then came the heat. In Oregon, temperatures reached a previously unimaginable 111 degrees on June 27. Places like Hood River, known for its windsurfing and pleasant year-round temperatures, felt more like the inside of a blast furnace in July. In Portland, where many houses don't have air conditioning because historically it was never needed, nearly one hundred residents died from heat stroke and other heat-related problems. In the previous twenty years, only two heat-related deaths were recorded in Portland and surrounding Multnomah County. In 2021 alone, extreme heat, drought, and wind combined to spark nearly fifty thousand wildfires nationwide, sending seven million acres and five thousand homes and other buildings up in smoke and costing taxpayers $4.5 billion in fire suppression costs—another record.[25] In Colorado, Denver experienced its first-ever autumn on record without snow, with no measurable accumulation from April to Thanksgiving. As the city prepared to celebrate the holidays, the probability of fire was higher than the probability of snow or rain. On Thanksgiving, the National Weather Service forecast had no signs of precipitation, but ominously predicted weather conditions that were "conducive to the rapid spread of new fires." One month later, the warnings of fire turned into horrific reality. Over two days at the end of December 2021, the most destructive fire in Colorado history ripped through suburban Boulder County, destroying more than one thousand homes and other buildings worth an estimated $1 billion or more.[26]

While fires raged once again in the West throughout 2021, in the East, Hurricane Ida in late August walloped Louisiana and other Southeast states, bringing with it floods and high winds and causing an estimated $18 billion worth of direct damage,[27] and billions more when you included lost productivity from the power being out for weeks in large swaths of the state. Ida would become the second-most expensive storm in state history, behind only Katrina, which coincidentally had hit the state exactly sixteen years earlier. Katrina had taken aim at downtown New Orleans, sending city dwellers scrambling to their rooftops to escape rising floodwaters and to signal passing helicopters for help. Ida, on the other hand, ripped up rural Louisiana. The state's agriculture industry lost more than $500 million in crops as sugarcane fields were soaked and stands of trees were pulled up by their roots, decimating the state's timber industry.

As it moved its way north, Ida showed that no segment of America—rich or poor, rural or urban, North or South, Republican or Democrat—can escape the economic costs of climate change. While in the Deep South Ida's damage was limited mainly to poorer rural areas, in the Mid-Atlantic and Northeast, it ravaged some of the most expensive parts of the nation's biggest cities. In Pennsylvania, rain and the rising Schuylkill and Delaware rivers caused an estimated $120 million in damages to roads, bridges, and other public infrastructure. It also flooded hundreds of homes, including million-dollar mansions and nearly every other building along historic Front Street in Bridgeport; shut down the SEPTA train line as well as the Vine Street Expressway and other main arteries in and out of Philadelphia's City Center; and rang up an estimated $100 million in damage across the state.[28] Then Ida moved on to New York City and New Jersey, where floodwaters filled basement apartments, trapping and drowning nearly fifty people, and poured into subway stations and Manhattan skyscrapers. By the time it finally stalled out over the

Gulf of St. Lawrence a few days later, it had churned up more than $65 billion in direct damage, making it the sixth-costliest Atlantic hurricane on record.

In 2021, another twenty disasters that cost $1 billion or more struck the United States. At a total pricetag of $145 billion, the damage they inflicted on the economy was greater than the gross state product of Mississippi, New Mexico or more than a dozen other states. The year's disasters began with a freakish February freeze in Texas that locked up natural gas pumps, shut down power plants, and knocked out the electricity across the Lone Star State. The year ended with the Marshall fire near Denver on December 30, preceded just weeks earlier by a rampage of rare December tornadoes that ripped across Kentucky, Illinois, Tennessee, and other states, killing about a hundred people and destroying an Amazon distribution center in Illinois, a candle factory in Kentucky, and other businesses along the way, just as they were trying to meet the Christmas rush and amid national supply chain problems. Meteorologists and climate scientists said record high temperatures across much of the South and Midwest in late 2021 added deadly warm fuel to the tornadoes, which typically occur with thunderstorms in the spring and summer, not the middle of winter.

Before the 2020s, the most billion-dollar disasters the United States had ever experienced in a single year was sixteen. In the decade between 2010 and 2020, there had been fewer than twelve billion-dollar disasters per year. In the decade before that? About six per year.

Just as we can now track the rise in global temperatures from climate change over time, we can also now track the growth in economic costs from climate-related disasters. The correlation is as clear as it is ominous. In the 1980s, when NOAA first started tracking the costs of the nation's biggest climate-related disasters, the average annual cost to our economy was about $18 billion. By the 2000s, it averaged about $54 billion per year. In the 2010s, it hit $85 billion. During the first two years of the 2020s, climate-related disasters averaged more than $120 billion per year. By comparison, the biggest climate-related weather disasters had an average annual impact of about $19 billion a year during the 1980s.[29]

Left unchecked, the economic damage from climate change is on track to only get worse. Much worse.

Climate-related disasters in the United States are projected to cost as much as $500 billion every year by the end of the century, five times what it was in 2021, according to the 2018 congressionally mandated National Climate Assessment.[30] According to the assessment, which included input from more than three hundred scientists, educators, business leaders, and others, those costs include $155 billion in annual lost wages from extreme heat and other climate-related issues; nearly $120 billion in damages each year to the nation's $1 trillion worth of coastal properties; and hundreds of billions more in damage to roads, bridges, and other infrastructure. Other estimates are even higher: An April 2021 study by insurer Swiss RE estimates that the world could lose 10 percent of its gross domestic product by 2050 due to unabated climate change. The United States alone could see its GDP decline by more than 9 percent by 2050—an amount equal to nearly $2 trillion[31]—as sectors ranging from real estate and tourism to agriculture and construction are hit hard by rising sea levels, droughts, flooding, and heat stress. The poorest parts of America will be hit hardest: Climate-related disasters are projected to reduce the economic output of the poorest U.S. counties by as much as 20 percent by the end of the century.

Of course, the United States isn't alone when it comes to climate change or the economic impacts from it. The global costs of climate change are racking up quite literally every day. According to a study by the World Meteorological Association (WMA), climate change is costing the world's economy about $383 million *every single day* on average, based on data from 2010 to 2019.[32] That's about seven times the amount of daily losses from climate-related events fifty years earlier. Add it all up, the WMA estimates, and the economic losses from climate-related events over the past fifty years is about $3.6 trillion—an amount bigger than the entire gross domestic product of the United Kingdom and twice the size of Canada's GDP.

But while every country is bearing the cost of a warming world and the climate change it's causing, the damage to the U.S. economy—borne by everyone from insurers to business and property owners to every American taxpayer—is growing faster than much of the rest of the world. That threatens to put the world's biggest, most powerful economy at an economic disadvantage to other countries. Globally, six of the ten costliest disasters in 2020 were in the United States, according to the insurance company Munich RE.[33] Insurers reported $74 billion in losses from U.S. disasters alone in 2020—five times the losses reported just ten years earlier, according to the Insurance Information Institute. Those costs are ultimately passed on to every American who has an insurance policy or owns or wants to own property. In October 2021, the average U.S. homeowner's premium was $1,312, according to the insurance institute, up 34 percent from a decade earlier and about five times the average premium cost in Europe. And those costs are going up. Following the 2020 fire season, more than fifty insurance companies in the nation's biggest insurance market, California, filed rate increase requests asking regulators to raise premiums for homeowners by as much as 40 percent.[34]

Along with increasing property losses, the rising costs of dealing with disasters in the United States also is exponentially large—and growing. In 2010, taxpayers in the United States spent about $809 million putting out wildfires, according to the National Interagency Fire Center. Just a decade later, taxpayers were shelling out nearly five times that amount. As of 2020, the United States employed about 370,000 firefighters. By comparison, that was three times the number of firefighters employed in all of China, and 50 percent more than all the countries in the European Union combined.

California's Dixie Fire, which in the summer and fall of 2021 torched almost a million acres and destroyed 1,300 homes, is a perfectly horrific illustration of these costs. Over the course of four months, Dixie became the single-biggest wildfire in California history and the most expensive fire suppression effort in U.S. history. During the month of August 2021 alone, the fire engulfed an area the size of the state of Rhode Island and almost five times as large as all of New York City. Entire forests of California redwoods, which typically live for thousands of years and were once thought to be impervious to fire, were killed by fires that burned hotter than anything they ever withstood before. To contain the conflagration and keep it out of more populous cities in Northern California, federal, state, and local governments sent nearly four thousand firefighters, 250 engines, and twenty-four helicopters into the battle throughout August. The cost: nearly $9 million every day, according to data from the National Interagency Fire Center. In all, taxpayers shelled out in excess of $610 million to deploy more than five thousand firefighters and their equipment before the nightmarish conflagration was finally contained just before Halloween, 2021.

The connection between increasing fires, hurricanes, floods, and a warming planet are clear. Now, so is the connection between global warming and rising economic costs. To see the trajectory of the economic costs of climate change, you don't have to study stock market closing prices or examine corporate balance sheets. Just look at the annual temperature rise in the United States and around the globe. Nineteen of the twenty warmest years have occurred since 2000. The year 2020 tied the hottest year on record, 2016 (which also happened to be one of the costliest years for natural disasters in America, according to NOAA). The summer of 2021 was the hottest summer on record, with the average temperatures in June through August about 2.6 degrees above the average for all of the twentieth century. Ominously, the record-tying temperatures in 2020 and 2021 occurred despite the presence of a La Nina effect that was supposed to make things cooler than average. Just as ominously, the Arctic showed that the Earth has a fever that keeps rising. The Arctic between 2020 and 2021 warmed twice as fast as the rest of the planet, with average temperatures reaching 10 degrees above normal at times. In June 2021, the temperature in Siberia hit a record 118 degrees. Including 2021, the last fifteen years saw the lowest fifteen extents of Arctic sea ice on record, according to the National Snow and Ice Data Center.[35] The icebergs are indeed melting.

But beginning in 2020, the discussion around climate change started to shift. Climate change was no longer just about melting icebergs or animals disappearing in faraway lands. It was no longer something measured only in degrees centigrade or parts per million. Climate change could now be measured in dollars and cents. It was hitting the pocketbooks of everyday Americans from one side of the country to the other, and all in between. And it was increasingly becoming a full-blown economic crisis that could no longer be dodged by deniers scoffing away the existence of global warming or suggesting it was a big hoax perpetrated by the Chinese. "We've seen this long freight train barreling down on us for decades," Michael Gerrard, director of the Sabin Center for Climate Change Law at Columbia University, told the *Los Angeles Times*, "and now the locomotive is on top of us, with no caboose in sight."[36]

There are two sides of the coin when it comes to the economics of climate change, however.

While climate change is killing our economy, the economic benefits of climate action could save it. Clean-energy, electric vehicles, and energy efficiency innovation are shaping up as the new cornerstones of our economy's foundation. They're creating new business opportunities worth billions and jobs in the millions. They're poised to reshape an economy battered by climate change, COVID-19, and inequity into one that's cleaner, more secure, more resilient, and more equitable.

Before renewable energy became mainstream, electricity production was the biggest generator of climate change–causing carbon emissions in America (in 2021, it was no. 2). Of all the new electricity brought online in America between 2020 and 2021, more than 70 percent came from the sun, wind, or other renewable energy sources—and all of it produced electricity more cheaply than fossil fuels. According to the investment firm Lazard, one megawatt of electricity in 2021 cost between $26 and $50 if it came from wind; $28 and $41 from utility-scale solar; $65 and $152 from coal; and $131 and $204 from nuclear. Only electricity from existing combined-cycle natural gas plants was—sometimes—competitive with utility-scale renewables, costing between $45 and $74 per megawatt hour.[37]

Most of the renewable energy projects in America were built in the last decade, attracting billions in investments and creating millions of jobs. Transportation displaced electricity production as the no. 1 generator of carbon emissions (it used to be no. 2).[38] Now, automakers, like utilities before them, are moving toward producing cleaner products. In 2020 and 2021, nearly every automaker announced it was either moving away from or completely abandoning the business of making vehicles that depend on oil and gas and shifting into high gear with electric vehicles, from Volkswagen and GM's

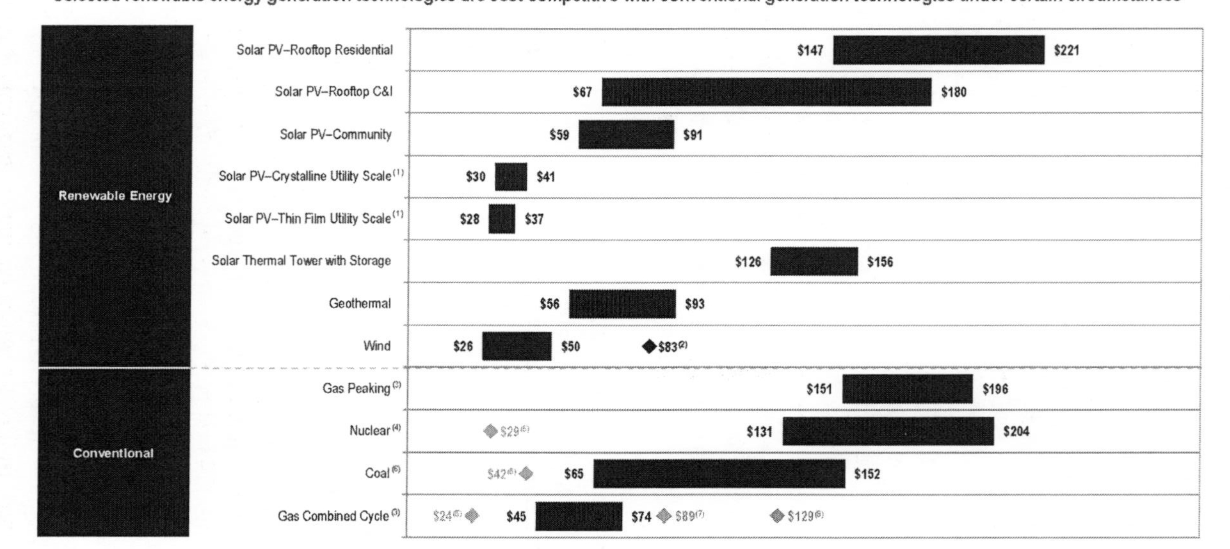

Figure 1.2. Wind and utility-scale solar were the cheapest sources of energy in 2021.
***Source*: Lazard, "Lazard Levelized Cost of Energy Analysis, v. 15 (October 2021)."**

plans for all-electric lineups to the electric version of the best-selling vehicle in America, the Ford F-150. Three decades after Detroit killed the electric car, it now sees EVs as the savior of the industry. Investors are grasping what's happening, too, and in turn are fueling the speed of change with their money. The third-biggest producer of carbon pollution in America is big industry, led by cement, iron, and steel production. In 2020, dozens of companies received major investments and launched companies to produce carbon-free cement and metals, and the leading research firm McKinsey & Co. optimistically predicted that cement companies could be on the path to reducing 75 percent of their emissions by 2050.[39] In keeping with that projection, more than forty of the biggest members of the Global Cement and Concrete Association announced in November 2021[40] that they would eliminate their carbon emissions completely by 2050.

Similar transformations are happening elsewhere throughout the economy. The nation's new and existing buildings (the no. 4 source of carbon) are increasingly being powered by renewable energy instead of gas. In 2020, more than fifty cities across America outlawed the use of gas in new construction,[41] while California—the nation's biggest housing market—also mandated that every new home come with solar panels. In December 2021, New York City banned gas hookups for heat, hot water, stoves and other appliances in most new buildings beginning as early as 2023. And in agriculture, the fifth-largest source of carbon emissions, low-carbon agricultural practices and products received a major boost in 2020 from federal and state programs designed to encourage such practices. In part because of these programs, private investments in ag-tech companies soared 137 percent in 2021 to more than $12 billion. Since 2017, investments in ag-tech companies have increased by 410 percent, according to analysis by *CropLife News*.[42]

Investors are beginning to respond to the shift to a cleaner economy in other ways as well. In 2020, the benchmark S&P 500 index rose by 16 percent, despite the COVID-19 economic downturn. That was pretty good. But it was nothing compared to returns in the clean-energy sector. The First Trust ISC Global Wind Energy Index Fund, which holds stock in about fifty U.S. and foreign wind companies, rose 81 percent. The Invesco Solar ETF rose by a whopping 234 percent.[43] On the other end of the energy stock spectrum, shares in the SPDR S&P Oil and Gas ETF, which holds about fifty U.S. and foreign petroleum company stocks, fell by 38 percent, while the iShares US Oil and Gas Exploration ETF sank 28 percent. Along with Wall Street, the financiers of the last great economic evolution in America—Silicon Valley tech investors—are investing in clean energy and so-called "climate tech" in a huge way too. In 2021, venture capital investments in climate-tech companies hit more than $31 billion, according to deal tracking firm PitchBook.

That was two-and-a-half times the amount invested in the sector just two years earlier—and it's only growing.[44]

The economics of climate change are also reshaping America's workforce. In 2020, more than three million Americans worked in clean energy—including renewables, energy efficiency, and clean vehicles.[45] That's more people than worked as schoolteachers, bankers, real estate agents, or farmers in America—and more than twice as many people as work in fossil fuels. Four times as many Americans worked in solar as worked in coal at the start of the decade. Eight times as many people worked in energy efficiency as worked in gas extraction. The shift to clean-energy jobs will only grow exponentially as today's young people enter the workforce. In a September 2020 poll, Morning Consult asked one thousand "Generation Z" Americans ages thirteen to twenty-three what kind of job they would want if they got into the energy business. About half said solar, and 43 percent said wind. Only 29 percent said natural gas, and 15 percent said coal. Other polls show that young people are increasingly putting purpose over paychecks when it comes to what motivates them to take a job. That's a shift. In the 1990s, young people graduating high school and college wanted to work in business and finance, the best-paying jobs of the time. In the 2000s it was computers and tech. In the 2020s, it is clean energy and jobs related to the environment.

Perhaps most importantly, the economics of climate change are doing what decades of environmental activism, scientists' warnings, and public protest could not do: It is finally moving politicians and policymakers into meaningful action. When politicians have thousands of clean-energy workers in their home state or district, it's easier for them to vote for legislation to advance clean-energy policies. It's also getting harder for politicians to deny climate change exists when wildfires, hurricanes, and floods continue to wipe out billions in home values, decimate small businesses, and make their states and regions less attractive to weather-weary big employers. Politicians can pretend that science doesn't matter or that the climate is changing with or without the impacts of mankind. But when climate change starts to hit constituents' pocketbooks, politicians pay attention, or they lose their jobs. Far beyond the borders of green-edged liberal states like California or Massachusetts, states including Nevada, New Mexico, North Carolina, and Illinois passed major climate policies in 2020 and 2021. Those policies are already only beginning to transform those states' economies in ways not seen since their evolution from farming, textile, and mining-based communities decades ago.

And then there's the federal government.

Not since Franklin Delano Roosevelt's New Deal has there been a bigger, more comprehensive attempt by the federal government to reshape and remake America's economy. The Biden administration's climate and clean energy agenda promises to catalyze massive amounts of investments in clean energy and clean transportation in every state in the nation, dramatically speeding up climate action in America and finally doing something to slow the economic costs of climate change.

When Joe Biden got started in politics in 1973, climate change was hardly an issue. Later, it became a third-rail issue, one that politicians avoided because it was so controversial. But when climate change clearly became a pocketbook issue with the start of the 2020s, it became a national priority. Wildfires, hurricanes, flooding, and drought are impacting every American who owns property, has insurance, or eats. Clean energy and clean vehicles are now creating real jobs, real companies, and driving real investments. You can see it in every state in every corner of the country, from electric vehicle factories in the Midwest and Southeast to wind turbine farms in the West and solar panels now just about everywhere. That wasn't the case in the past.

For the first time in U.S. political history, climate change in 2020 became a core issue during a presidential election. Nobody staked their career, their political future, and the future of the country on climate action like Biden. As president, he has pledged to cut America's greenhouse gas emissions by 50 to 52 percent by 2030; source 100 percent of America's electricity from solar, wind, and other clean-energy sources by 2035; and convert our transportation system to all-electric and other zero-emission vehicles by 2050. Those are monumental, game-changing goals for our economy, the likes of which we have never before seen. If they're reached, it means anybody born in the year Biden was elected will probably generate their own electricity or get it from solar and wind power companies by the time they're old enough to have to pay their own power bill. By the time they start their own families, their new minivan or sedan will probably be powered by electricity or some other net-zero emission fuel. And by the time they're ready to retire, the United States could be removing more greenhouse gas emissions from the atmosphere than it puts into it. If scientists are right, that means global temperatures will subside, and with them, the impacts of climate change. If economists are right, clean-energy jobs, innovations, and investments will increase dramatically.

Of course, presidents from Lyndon Johnson to Barack Obama have tried to pass policies to address climate change in a meaningful way. All of them failed. What makes this time different, though, is the economics of climate change—climatenomics. Biden and his cabinet of climate experts have seized upon that fact to turn *climate* policy into *economic* policy. They know that the key to lasting and enduring climate policy is not talking about polar bears or melting ice caps or even health and the future of our children. They know the key is to talk about jobs and the economy. "When I think about climate change, I think about jobs," Biden said during the 2020 campaign and repeatedly since. Polling shows that Americans are finally beginning to make the connection between the environment and the economy, and they want the government to do more to address the economic costs of climate change and expand the economic benefits of climate action.

To be sure, Biden's Republican opponents remain steadfastly opposed to anything the president does. And the long and tortuous fight over his climate and clean energy legislation in Congress showed that not even Democrats are united when it comes to paying the high costs of investing in clean energy and climate action. There also are certainly no guarantees that the promises Biden made to the American public to transition the nation to a clean economy, or the pledges he made to world leaders that America will lead the world in cutting greenhouse gas emissions, will come to fruition.

Regardless of what happens with Biden's climate and clean-energy agenda, however, the die is cast. Businesses are already reacting to the market signals being sent by lawmakers in Washington and by governors and statehouses across the country who also are regularly rolling out new climate polices. Investors are moving big money from yesterday's business behemoths like Exxon to today and tomorrow's companies like Tesla and Sunrun. Together, the potent and proven combination of business and government, coupled with Americans' desire for climate action and workers' desire for good-paying jobs that will last, is poised to fundamentally reshape America's economy in ways we haven't seen in generations.

But the transformation now underway in response to the economics of climate change also will do much more. It promises better prepare our communities for tomorrow's weather disasters and bolster our economy amid the rising costs of climate change. It promises to improve equity in America by directing clean energy investments into communities of color and making electric vehicles, solar panels and more efficient appliances and homes accessible not just to the rich, but to low- and moderate-income Americans as well. It promises to improve our national security by reducing our dependence on foreign oil, and reestablish America's leadership on the existential global issue of our time.

And with a little luck, it might just save our planet.

Chapter Two

Drowning Chickens, Soaked Marines, Bomb Cyclones, and Paradise Lost

There was never anything easy for Easy Company.

The unit of the 2nd Battalion, 2nd Marines was forged of sand, seawater, and enemy fire on the beaches of Guadalcanal in the first major World War II campaign against Japan. A year later, the unit spearheaded the assault on Betio during the battle of Tarawa, where 180 Easy Company Marines hit the beach on November 20, 1943. Only forty walked off the island on their own.[1] Marines from Easy Company would go on to see battle in Afghanistan and Iraq, where they were among the first to fight in Fallujah.

Casualties aside, in terms of damage and destruction, little could compare to what Easy Company and the rest of the Marine Corps stationed at Camp Lejeune, North Carolina, experienced in September 2018, when Hurricane Florence stalled over the base, pummeling it like an unreachable enemy no weapon could stop.

In war and peace, heavy weather was nothing new to the Marines, of course. Two years earlier, at the other end of the Carolinas, staff and recruits at Marine Corps Recruit Depot Parris Island had to evacuate the base in advance of Hurricane Matthew.[2] As Florence churned her way toward the Carolinas, personnel at Parris Island were once again in full retreat to higher ground. Company by company, several thousand Marines and recruits boarded up windows and then boarded a convoy of buses that would carry them inland, rolling out in orderly procession under darkening skies.[3] To the instructors and other staff, the maneuvers required in the face of flooding had become nearly as methodic as a flanking tactic or pincer movement.

At Camp Lejeune in 2018, the story was different. There, base commander Brigadier Gen. Julian D. Alford made the decision for his more battle-hardened Marines to stay put in the fight against Florence. Alford knew about weather along the coastal Southeast. He grew up in Atlanta.[4] As a young captain, he served as a company commander at Parris Island and earned his first star as commanding general of the Marine Corps Warfighting Laboratory/Futures Directorate in Quantico. Battle-hardened and time-tested, he had seen combat in Panama, Iraq, and Afghanistan and had earned a reputation for his dedication to the Marines and his ability to outwit the enemy. Florence, Alford figured, was no match for the Marines. Amid rising criticism from worried military families and personnel,[5] Alford defended his decision for the Marines to stay put at Lejeune, saying the base was well-prepared for bad weather, with emergency shelters, backup generators, and hardened infrastructure and buildings that had stood the test of time. "Camp Lejeune has endured countless destructive weather events over its seventy-seven-year history, and we will withstand the tough conditions ahead," Alford said.[6]

But Alford had no idea what was about to hit him and his Marines. In the age of climate change, history no longer mattered. In the course of just twenty-four hours, more than twenty-five inches of rain fell on Camp Lejeune and neighboring Jacksonville. Over the next few days, another foot of rain would fall. Then came more rain, not just at Camp Lejeune, but across every corner of the Carolinas. By the time slow-moving Florence, stretching four hundred miles from end to end, finally moved on, an estimated 9.6 trillion gallons of rain would fall on North Carolina, making it the worst disaster of its kind ever recorded in the state.[7] Locals whose families had seen coastal storms and hurricanes for generations had never seen or heard of anything like this, nor had anybody else for that matter. More than fifty people were killed by the storm, carried away from their homes and cars, drowned by flash flooding, or crushed by falling trees made vulnerable by high winds and water-logged roots. In Union County, two hundred miles inland from the North Carolina coast, the body of a one-year-old boy was found in the back-seat of a car swept away by floodwaters. His mother was unable to wrest him from the vehicle before he was washed away from her forever.[8] The initial flooding from Florence was just the beginning. As water that fell on the North Carolina mountains and highlands flowed east to the sea, rivers filled with more water than they had ever handled easily crested their banks. So did uncovered farm lagoons filled with hog and poultry feces, sending millions of gallons of excrement flowing into neighborhood streets, making the air unbreathable, the water undrinkable, and homes unlivable.

With such widespread destruction, from the mountains to the sea, it would take months before the state of North Carolina could even get a good estimate on the amount of damage Florence wrought. When the numbers did come in, it was staggering. At least $17 billion in damage was caused by the storm, not so much from high winds—Florence was only a Category 1 storm by the time it made landfall—but from flooding.[9] From the three-hundred-year-old city of New Bern near the coast to the college town of Chapel Hill in the middle of the state to the slopes of Sylva and other parts of the North Carolina mountains, swollen rivers and creeks sent water rising up the walls of offices, homes, and historic buildings, leaving behind mold, mildew, rotting drywall, and dead animals and crops as it receded. In the little town of Trenton, east of the Croatan National Forest, there were so many leaking roofs after relentless days of rain that the town and neighboring communities ran out of tarps to cover the holes. Six months later, still-closed shops, offices, homes, and churches along Jones Street, the main drag through town, shared an odd characteristic: Inside almost every building, the bottom-half of the interior wall was missing. Their owners had cut out the soggy material above the flood line to prevent mold and rot.

Back at Camp Lejeune, the damage assessment was hard to fathom. Of the 6,182 military housing units at Camp Lejeune, 70 percent of them—more than 4,300 homes and apartments—were damaged by Hurricane Florence and the floods that followed.[10] Of the thirty-one other military buildings on the base, 100 percent were damaged, many so badly the military would later deem them not worth saving. Roads were washed out, roofs ripped to shreds, and trees split and splintered as if by artillery fire. In all, rebuilding Camp Lejeune was estimated to cost more than $3.6 billion—an amount equal to almost 10 percent of the entire annual budget for the Marine Corps. At a December 2018 Senate Armed Services Subcommittee hearing, Secretary of the Navy Richard Spencer warned lawmakers that Camp Lejeune wasn't alone. Military bases are "exposed to what we've seen now with 100-year storms that come every two or three years," he testified. Even as his boss, President Donald Trump, mocked the idea of climate change and dismissed its impacts, Spencer told lawmakers, "We're going to have to start addressing this."[11]

In fact, just one month after Florence, Hurricane Michael slammed into the panhandle of Florida, and in doing so, nearly wiped Tyndall Air Force and the 325th Fighter Wing off the map. About eleven thousand people connected to the base had to be evacuated. The damage was just as bad as, if not worse than, it had been at Lejeune a month earlier. At Tyndall, an estimated 95 percent of all buildings and other infrastructure were damaged by the hurricane. The Air Force estimated it would cost $750 million to repair or replace buildings and other structures and make the base whole again. That didn't include the cost of repair and replacement of aircraft, including at least some of the two squadrons of F-22 Raptors—the world's most expensive airplanes, costing taxpayers nearly $350 million each—that were stationed at Tyndall. A year after Michael, Tyndall was still rebuilding. Base officials said it would take five years before the base was fully mission-ready again.

Five months after Michael, another military base in another part of the country found itself in the crosshairs of climate change. On March 13, 2019, a massive storm stalled over the Plains states, setting low-pressure records in four different locations that left meteorologists bewildered. Winter Storm Ulmer would become a "bomb cyclone," unleashing a deluge of biblical proportions across the Midwest. No place was hit as hard as Offutt Air Force Base in Nebraska. Opened in 1921, Offutt is one of America's oldest military installations. During World War II, workers there built the Enola Gay and Bockscar, the B-29 bombers that dropped the atomic bombs on Japan that ended the war. Offutt would later become home to the U.S. Strategic Command and, coincidentally enough, the home of the Air Force Weather Agency. It also is home to the 55th Security Forces Squadron, which is responsible for helping defend America's nuclear arsenal as well as maintaining and operating the "Doomsday Planes," a squadron of four $223-million militarized Boeing 700s designed to replace the White House and serve as a flying command center for the president in case of nuclear war or other national emergency. Despite its long history, Offutt had never experienced anything like it did during March of 2019. Floodwaters washed out roads, making them impassable and cutting off personnel from the headquarters of the 55th. Even if airmen and women could get to the headquarters building, it's unclear they could've launched a Doomsday Plane or any other aircraft from Offutt if they had to: The sole hangar housing an undisclosed number of the Doomsday Planes was damaged, flooded, and unreachable, as was about three thousand feet of runway.[12] In addition to the flight line operations, the historic flood destroyed more than 130 other buildings at Offutt. A month later, even more buildings were deemed unusable because they were covered in dangerous black mold. Six months after the storm, the sneak attack of flooding and mold at Offutt still had much of the base out of operation.[13]

Shortly after the flooding at Offutt, the Air Force delivered the bill for taxpayers for the disasters at Tyndall and Offutt in the form of a request to Congress for supplemental funding. The cost of climate change for those two bases alone: Nearly $5 billion.[14] Air Force Secretary Heather Wilson warned it would likely be even higher as damage estimates from Offutt were completed. At a speech at the conservative Heritage Foundation, which held great sway with President Trump and Republicans in Congress and had spent years characterizing climate change and its economic costs as left-wing drivel, Wilson warned that without more money from Congress, the Air Force budget would be so strapped it would impact its ability to protect the country. A former Republican congresswoman who represented New Mexico's 1st District for a decade, Wilson understood politics and the budget process well. Not prone to hyperbole, she made her point clearly to the conservatives at the Heritage Foundation. "If we don't get a supplemental (funding bill)," said Wilson, "it's going to affect the rest of the Air Force and our ability to operate."[15]

The disasters at Camp Lejeune, Tyndall, Offutt, and other military operations around the world illustrate just some of the growing economic and national security costs that come with climate change—costs that are borne by every American taxpayer. But the costs of climate change for the military go much deeper than just the immediate expenses associated with cleanup and rebuilding after storms. The military has been battling the impacts of climate change—and taxpayers have been paying the cost of it—for decades. Climate-related costs have become a fast-growing part of the operating costs for the military in every part of the world, from the polar ice caps to the deserts of the Middle East. And they're only getting higher every time the globe's temperature rises.

For at least forty years, the military and climate scientists have watched the ice around the North Pole recede, opening up new shipping and attack lanes between Russia and the United States and Canada. As a result, the United States in recent years has been forced to spend billions of dollars in the region on new radar and sensors. America will need to spend billions more soon to protect the country because of the changing climate in the northernmost part of the world. One simple but expensive example: As of 2022, Russia had a fleet of at least forty icebreakers that could operate in the Arctic. The United States had two.[16] Just to keep up with the Russians, the Coast Guard estimates it needs many more of the ships, which cost as much as $1 billion each. Climate change also is posing new threats—and costs—for the military as it is forced to engage in an increasing number of humanitarian and relief efforts around the globe. Look to the Middle East to see how that's playing out. There, steadily rising global temperatures have resulted in record droughts and water scarcity that have decimated the agriculture industry in Syria and other parts of the Middle East. That economic impact of climate change in turn helped trigger unrest that led to the 2011 Arab Spring uprising and civil war in Syria, studies show.[17] In the years that followed, the U.S. military's involvement in Syria and Iraq" cost American taxpayers more than $23.5 billion, Defense Department records show.

In October 2014, then-U.S. Defense Secretary Chuck Hagel, a former Republican senator, laid it out plainly during a speech in Peru.[18] "Climate change is a threat multiplier," he said. "It has the potential to exacerbate many of the challenges we already confront today—from infectious disease to armed insurgencies—and to produce new challenges in the future," Hagel said at the Conference of Defense Ministers of the Americas. "The loss of glaciers will strain water supplies in several areas of our hemisphere. Destruction and devastation from hurricanes can sow the seeds for instability. Droughts and crop failures can leave millions of people without any lifeline, and trigger waves of mass migration." In connection with Hagel's 2014 speech, the Department of Defense released a first-of-its-kind Climate Change Adaptation Roadmap.[19] The plan included steps to measure and address the costs of national security vulnerabilities from melting ice and rising seas, the impacts of increased global instability from extreme weather and the myriad problems from sea-level rise, extreme weather, flooding, and other climate-related disasters facing the military's more than seven thousand bases and other installations worldwide. Though the threats to military operations have continued to rise dramatically since then, the preparations have not. After President Trump took office, he basically abandoned all climate change preparations by the United States. In a March 28, 2017, executive order, he rescinded climate-related reports and actions, including Hagel's previous directives at the DOD.[20]

In some cases, growing costs of climate change to the U.S military and taxpayers who pay the military's bills are simply too much to even fathom. The truth is, the government just doesn't know what to do in some cases. Consider Naval Station Norfolk, headquarters for the U.S. Navy's operations in the Atlantic, Mediterranean, and Indian Ocean, and the world's biggest naval base, with seventy-five ships and more than 130 aircraft that take off on average every six minutes. The base covers more than 3,400 acres, including eleven piers stretching across four miles of Atlantic waterfront.[21]

Since the base opened in 1917, the sea level along Naval Station Norfolk has risen about fifteen inches—enough to cause millions of dollars' worth of damage in recent years.[22] Massive concrete and wood piers that held and serviced Navy ships for decades in recent years were submerged under rising seas, making them unusable for sailors and service contractors and the ships that needed to tie up to the docks. At least ten times a year, the base faces major flooding that closes roads to passage, shuts down aircraft runways, and impacts ship operations. By the year 2100, the base will likely flood 280 times per year—an average of more than five times per week—according to a Union of Concerned Scientists report.[23] Repairs and retrofits to address sea-level rise and other climate-related problems at Naval Station Norfolk have already cost billions of dollars, impacting American taxpayers every time Congress approves the annual appropriations for the Department of Defense. Military officials say they need to replace more piers and other infrastructure just to ensure the base can continue to properly dock and service the ships based there. The costs could be so high that it would literally be financially unfeasible, making America's largest Navy base not worth the money required to save it.

"Sea-level rise and flooding will be one of the factors considered in evaluating the military value of the base," Navy Capt. Joe Bouchard, who from 2000 to 2003 was commanding officer of Naval Station Norfolk, told the *Virginian-Pilot* newspaper.[24]

The problem is, if America doesn't invest billions of dollars into Naval Station Norfolk to save it, what should be done with the base instead? You can't just pick up the world's largest Navy installation and all the operations and personnel it supports and move it inland any easier than you could pick up the Pentagon and move it a few miles down Interstate 95. Military leaders are perplexed. For the Norfolk–Hampton Roads region, a base closure would be devastating. About 40 percent of the region's economy is tied to Naval Station Norfolk. But even if the base goes away, the impacts of climate change and the costs that come with it won't. A 2016 study by the College of William & Mary Law School's Virginia Coastal Policy Center warned that sea-level rise could cost tens of millions of dollars each year if the region doesn't do anything to address coastal flooding.[25] The quicker the ocean rises, the quicker and higher the cost rises. If the sea level of the Atlantic at Norfolk rose by another foot and a half it would result in $50 million in annual damage from increased flooding, according to the report. If it rose by 2.5 feet, it would cost the region $100 million a year.

Homeowners in the Norfolk area are already feeling the costs as flooding and climate change take a toll on housing valuations. A September 2018 study by the nonprofit First Street Foundation showed that the value of homes in Newport News had declined more than $4 million since 2005.[26] In 2021, the Norfolk area suffered nearly $35 million in damage from flooding, according to First Street. Within thirty years, that is expected to increase by 70 percent, to about $59 million. To keep residents and businesses from leaving, cities in the Norfolk region are trying to do more to ease the impacts of climate change and the costs that come with them. But that costs money, too. In December 2020, the city of Hampton issued $12 million in bonds to help pay for three flood control projects. Hampton Mayor Donnie Tuck said taking on the debt was necessary to protect the region's future, thanks to climate change. "We know now that Hampton Roads is a leader in a less desirable category, one of the country's most vulnerable areas to sea-level rise and climate change," he said at a press conference. While coastal Virginia may hold a dubious position as one of the leading areas for flooding and flood-related damage, homeowners, businesses and government installations along every coastline are looking at an expensive future. A February 2022 NOAA study estimates that sea levels around the country will rise by a foot by 2050—increasing as much in the next thirty years as they did in the previous century.[27]

As with the military, the economic costs of climate change are adding up for another huge and important sector of America's economy, the agriculture industry. Years of climate-change-induced disasters, from the bomb cyclones in the Midwest to hurricanes in the East to drought and fire in the West, have devastated America's farmers. In 2019, farm income hit $88 billion—the highest level since 2014. But a closer look reveals a much less rosy picture. According to the American Farm Bureau Federation, about 40 percent of all farm income came from disaster assistance and other federal and state subsidies in 2019.[28] In 2020, farm debt hit a near-record high of $425 billion, despite federal assistance from COVID-19-related relief and other programs that all taxpayers will bear. Taxpayers sent about $78 billion in aid to farmers in 2020, including $46 billion in direct payments and $32 billion in pandemic and emergency aid programs.[29]

The bills are coming due for all Americans with every grocery charge, restaurant check, and income tax bill. Between 2020 and 2022, amid record flooding in the Midwest, hurricanes in the East and drought and fires in the West, the Bloomberg Agriculture Spot Index rose nearly 75 percent to record highs after weather disasters decimated corn, soybeans, wheat and other commodities. Yes, the general rise of inflation pulled up food and commodities costs just like they did the costs of cars and homes and just about everything else. But when it came to prices for crops and food, inflation was also being driven by climate change. Temperatures in the prairie states hit their highest levels ever during the summer of 2021, breaking records set during the Dust Bowl and pushing corn prices up 45 percent and wheat prices to their highest levels in years, according to *The Atlantic*. The magazine in a February 2022 story called climate's inflationary impact "The Rise of Greenflation." The *Wall Street Journal* two months earlier put it even more succinctly. Americans, the Journal wrote, could "Blame Bad Weather for Your Bigger Bills."

Along with paying more for food, American taxpayers are paying more to support the farmers who produce it, thanks in part to the economic impacts of climate change. Over the past decade, three main federal farm support programs managed by the U.S. Department of Agriculture—the Federal Crop Insurance Corporation, the Agriculture Risk Coverage Program, and the Price Loss Coverage Program—cost an average of $12 billion annually, depending on the extent of weather disasters each year. But in 2020, costs for farm subsidies was nearly four times that amount, including COVID-19 relief and other emergency aid.[30] Farmers pay annual premiums for crop insurance, but the majority of the funding for the federal crop insurance program comes from other taxpayers. So every time an acre of land gets wiped out from climate-related storms, it costs all of us, and those costs are rising. The record storms of 2018 and 2019 were so costly, in fact, that the USDA and Congress had to create a special new subsidy program to help offset farm losses that couldn't be covered by the federal crop insurance program. In June 2019, Congress authorized and President Trump signed into law the USDA's Wildfires and Hurricanes Indemnity Program Plus, known as WHIP+, setting aside an additional $3 billion for farm losses specifically from the rash of hurricanes and wildfires in 2018 and 2019.[31] Six months later, another $1.5 billion in taxpayer money went to the program. And that was before the record hurricanes, wildfires, and flooding of 2020 even began. Climate change is clearly driving up the costs of crop insurance. An August 2021 Stanford University study found long-term warming and drought contributed $27 billion to the losses covered by the U.S. crop insurance program between 1991 and 2017. Meanwhile, a USDA study predicts what's still to come. According to USDA researchers, climate change could increase the costs of the Federal Crop Insurance Program by as much as 37 percent by 2080 if we don't reduce greenhouse gas emissions.[32]

Scientists and farmers alike only have to look at what happened in Iowa in August 2020 to envision what the future could look like on the ground. The hurricane-force derecho that swept through Iowa then with winds up to 140 miles per hour and rains that pounded the state like water from a firehose wiped out about 850,000 acres of crops. More than 40 percent of Iowa's corn and soybean crop, which annually generates about $14 billion each year, was destroyed. The direct costs of the storm, estimated at about $7.5 billion, made it the most expensive thunderstorm-related disaster in U.S. history. But the immediate costs are just the beginning.[33] Increases in federal crop insurance, food costs, and the lingering financial impacts of the storm to farmers in Iowa and throughout the Midwest will last long after the storm is forgotten.

It's not just farmers in the nation's breadbasket who are hurting from climate change. While Iowa is the heart of the corn belt, North Carolina produces more poultry, tobacco, and sweet potatoes than any other state. It also is the no. 2 state (behind Iowa) for hogs and no. 6 for cotton.[34] In 2018, the same hurricane-related flooding that damaged so much of Camp Lejeune also destroyed an estimated $2.4 billion in crops, pigs, and poultry across the rest of the state. In Sampson County, east of Raleigh, the Burch family has grown sweet potatoes, collard greens, and other crops since the 1700s. After Hurricane Florence in 2018, patriarch Jimmy Burch walked his barren fields and surveyed what was left after the flooding that was unlike anything his family—or anyone else in North Carolina—had ever experienced. He estimated he lost about a third of his sweet potato crop when the storm hit shortly after planting season. "It was over a million dollars washed down to the Atlantic Ocean," he told *Marketplace* radio.[35] "It'll take me four or five years to make it back." Burch was one of the lucky ones. Other farmers lost 100 percent of their crop. Florence came just as many of the state's crops were about ready for harvest, and as much of the state's poultry, especially Thanksgiving turkeys, were being fattened up for market. Just two years earlier, Hurricane Matthew caused $400 million in crop and livestock losses. Many North Carolina farmers had been counting on the 2018 harvest to help them get back on solid financial footing. Then came Florence and more flooding. "We knew the losses would be significant because it was harvest time for so many of our major crops," said Agriculture Commissioner Steve Troxler.[36]

While farmers like Burch awoke in the days after Florence to flooded fields and lost crops, others woke to see millions of dead chickens, pigs, and turkeys, bloated and floating in the dirty floodwaters and rotting in mud and muck. An estimated four million chickens and turkeys drowned over the course of a few days, along with approximately 5,500 hogs. Sanderson Farms, one of the nation's biggest poultry producers, wasn't new to climate disasters. In 2005, Hurricane Katrina hammered its operations in Mississippi, where the company is based. But what Sanderson experienced in North Carolina was worse. There, rising waters flooded 70 of the 880 broiler houses at farms that contracted with Sanderson, the company's chief veterinarian, Philip Stayer, reported in Florence's wake.[37] Approximately 2.1 million chickens contracted to Sanderson drowned as water rose above their cages, trapping them as if in a sinking ship with the hatches locked. Even if the dead birds were salvageable, the company couldn't do anything with them if they wanted to. Processing plants across North Carolina were idled for days because of flooded-out roads, lack of electricity, and lack of workers who were busy cleaning up their homes and communities from the mess.

After the rains finally ended, Sanderson and other poultry growers faced new problems. Temperatures that reached above 90 degrees threatened the health of remaining live birds, especially in places that still didn't have power and couldn't run cooling fans essential to most hen houses. And then there was the problem of what to do with the carnage that remained. Fields were too flooded and water tables too high to bury the birds. So instead, farmers that worked for Sanderson and other poultry producers essentially had to let the dead birds rot in place. They first covered the floors of chicken houses with sawdust and other materials, letting the putrid hot stew of rotting flesh and feathers dry for up to two weeks. After that, farmers carried mounds of rotten birds out with shovels and front-end loaders, piling them up outside the houses and then covering them again with more sawdust as they decomposed. The state of North Carolina spent more than $11 million[38] on sawdust and other materials just to help struggling farmers compost the estimated four million turkeys and chickens lost in the floods, money that would eventually be reimbursed by the federal government, and in turn, taxpayers across the country.

The cost to taxpayers and the American economy from the flooding in farm states will remain long after the last chickens rot and the final soggy acres of corn are plowed under. If there is a bright spot, it's this: As climate change has shifted from an environmental issue to an economic issue, so has the sentiment and sense of urgency in some rural states. Look again to North Carolina. In 2010, a panel of scientists at the North Carolina Coastal Resources Commission forecasted sea levels could rise by thirty-nine inches in the next century, about the same level that waters at Camp Lejeune and other coastal communities rose in the wake of Hurricane Florence just eight years later. But back in 2010, North Carolina lawmakers were concerned less with protecting the future of their state and more with politics and short-term profits. Talk of sea-level rise and climate change, they knew, could put a damper on the state's coastal real estate market and keep tourists and homebuyers away. So the legislature basically outlawed sea-level rise forecasts when considering matters impacting the state's coast and coastal real estate.[39] Despite the international ridicule that followed, North Carolina lawmakers stood by their move, heads planted firmly in the proverbial sand, even as it washed away from the beaches around them. Not to be outdone, a few years later in the town of Woodland, local leaders placed a moratorium on solar panels. The reason? Townspeople, influenced by clean-energy opponents backed by the fossil fuel industry, said they were worried that too many solar panels could suck up all the sun's energy.[40]

After Michael and Florence and Matthew and other storms in between, however, the conversations about climate change began to change. "We've weathered two so-called five-hundred-year floods in two years, and three in fewer than twenty years," North Carolina Gov. Roy Cooper would tell a U.S. Senate panel in February 2019 as his state was drying out after Florence.[41] Cooper would go on to direct the secretary of the state's Department of Environmental Quality, Michael Regan, to create a major climate and clean-energy plan, which the state launched shortly before Regan was tapped by President Biden to become secretary of the U.S. Environmental Protection Agency. "Just like many places in our country and across our globe," Cooper said, "we're beginning to feel the harsh effects of climate change on our communities and our economy."

No place has felt the harsh effects of climate change on its communities and its economy more than the state with the nation's biggest economy and population, California.

Ironically enough, it was a human-influenced climate disaster of another generation, and the economic cratering that followed, that led to the rise of modern-day California. Like today's climate disasters, the Dust Bowl of the 1930s resulted from the combination of nature's whims and man's disrespect for them while in the pursuit of short-term profits at the expense of long-term sustainability. In the 1920s, farmers in the Great Plains states seized on the advent of modern farm machinery with reckless disregard, using new gas-powered tractors like Henry Ford's mass-produced Fordson, combine harvesters, and the revolutionary three-point hitch to plow away prairie grasses and dig deeper into virgin topsoil built over a millennium. The extended drought of the early 1930s would have been enough for farmers to deal with in itself; the drought combined with the self-inflicted wounds from farmers' wanton tilling of fertile topsoil in the Great Plains was too much. An estimated one hundred million acres of soil eroded and literally blew away in the wind across a swath of America stretching from Colorado and New Mexico to Texas and Missouri, with Oklahoma at its center. At one point in 1936, farm losses reached as high as $25 million per day, further ripping into the economic foundation of a country struggling through the Great Depression like bitter winds against a battered barn.[42]

Unable to grow a thing on previously lush farmlands that had turned to dust, hundreds of thousands of farmers and others wrapped their faces to protect themselves from grit and wind, loaded up their trucks, and uprooted their families in a desperate search for better life and greener pastures. More than four hundred thousand of these American climate refugees headed west to California, lured by stories about cool water that ran freely across green fields and endless sunny days unimpeded by dust storms. Old-timers remembered the stories they had heard about California back in their childhood, when Gold Rush advertisements promised streams filled with precious metals that were as plentiful as trout, and the California Promotion Committee—a predecessor to the state chamber of commerce—described the state as a "Cornucopia of the World." The Union Pacific Railroad lured tourists with ads showing young women amid bountiful valleys, snow-capped mountains, and warm beaches and the alluring tagline "California Calls You." Later, the man who would become the closest thing to a patron saint for Dust Bowl refugees, troubadour Woody Guthrie, serenaded his fellow Okies in his 1940 song *Do Re Mi* with descriptions of California as a garden of Eden, as paradise, as long as they could pay for it, that is.

Less than eight decades after Guthrie sang about California as paradise, Paradise turned into hell, and the question of whether the world has enough dough to pay for the economic costs of climate change came into stark and horrific reality.

Over the course of seventeen harrowing days in November 2018, the nation's worst wildfire in recorded history ravaged the tree-covered foothills of the Sierra Nevada that naturalist John Muir once waxed poetic about, turning it into a blackened horror scene of devastation. At the center of the conflagration was the little town of Paradise. Established in the 1860s during Muir's time, by 2018, Paradise had become a haven for retirees and big-city transplants looking for a quiet piece of California where they could raise their families and slow down below the shadows of the Sierras instead of the skyscrapers of San Francisco or Los Angeles. Paradise had its share of brushes with disastrous wildfires before. Ten years earlier, firefighters from a dozen states and the National Guard battled more than three hundred wildfires burning across Northern California simultaneously, engulfing the region in a deep blanket of smoke so bad that residents couldn't go outside without a mask. One of them, the Humboldt Fire of June 2008, came so close to the southern end of Paradise that nine thousand residents were forced to flee.[43] In July 2008, it was the northern end of Paradise staring down flames, when another fire once again forced thousands to evacuate.

But nothing compared to the horror to come a decade later.

Around 6:30 on the morning of November 8, 2018, firefighters were dispatched to a small brush fire off Camp Creek Road near the town of Pulga, about twenty-five miles east of Paradise. Driven by hurricane-force winds and fueled by bone-dry brush and trees, the Camp Fire, named after the road where it began, was at the doorsteps of Paradise within a couple of hours.[44] By day's end, more than thirteen thousand homes and five thousand other buildings in Paradise were destroyed. Seventeen days later, when the fire was finally contained, more than 153,000 acres were burned and eighty-five people were dead. Those who survived were left with the emotional and financial aftermath: More than 18,800 homes and structures worth $16.5 billion were destroyed, making it the world's costliest disaster in 2018.[45] About a fourth of those homes weren't insured, but in the new age of climatenomics, even insurance companies aren't a guarantee of anything. In the town of Atwater, three hours south of Paradise, the predecessor to the Merced Property and Casualty Insurance company was started in 1906 by farmers looking to insure their homes and farms from the occasional fire. After 112 years of protecting the region's residents from natural disasters, the company in December 2018 filed for bankruptcy, becoming another victim of the Camp Fire.[46] Merced Property and Casualty listed assets of about $23 million and liabilities of nearly three times that much, mainly from losses in the city of Paradise.

The demise of Merced Property and Casualty is emblematic of the financial tremors rumbling in the insurance industry with every climate disaster, threatening in no uncertain terms to rock the world's financial markets, and all of us who depend on property insurance, retirement savings, and lending markets. Insurance company payouts for natural disasters hit $219 billion during the years 2017 and 2018, the highest ever for a two-year period, according to reinsurance company Swiss RE.[47] As a result, insurers are raising rates, reducing coverage, and simply pulling out of some parts of the United States.

Nationwide, insurance companies raised rates by 50 percent between 2005 and 2015 in large part because of increasing natural disasters, according to the National Association of Insurance Commissioners.[48] In the fourth quarter of 2021, insurance premiums rose for the seventeenth consecutive quarter.[49] Coastal states hard hit by hurricanes, storms, and fires are feeling the broadest repercussions from the climate-related impact to the insurance industry, but no part of the country is exempt. The biggest premium increases in recent years, in fact, have actually been in tornado-alley states like Kansas and Oklahoma, where more intense storms and tornadoes drove up annual premiums by more than $650 on average.[50]

In California, the state insurance department in 2018 received sixty-nine rate increase filings from property and casualty companies, up from just twenty-five such requests a few years earlier. Even before the Camp Fire, California businesses and consumers in 2018 saw a 50 percent increase on average for insurance. And that was just for those lucky enough to still have insurance. After the Camp Fire, numerous insurance companies pulled out of the market, leaving homeowners and businesses struggling to find coverage and some involuntarily eschewing coverage altogether because of the high costs—particularly in parts of the state that are most vulnerable to wildfire, the so-called "wildland-urban interface areas." At the end of 2021, as many as twenty-four million homes in California were at risk of losing their insurance as state-mandated grace periods for homeowners insurance were set to expire.[51] "We're going to pay the bill for climate change one way or the other, and it's just a question of how we divvy up that cost," David Russell, codirector of the Center for Risk Management and Insurance at California State University, Northridge told *Bloomberg News*.[52]

As the state's former insurance commissioner, Dave Jones has been warning about these growing financial risks for years. "I think we're steadily marching toward a future where the (wildland-urban interface areas) are uninsurable," he told a group of climate experts and business leaders during Climate Week in New York in September 2019. "We are seeing global temperatures rising; we're not doing enough action fast enough to address global warming . . . and when the risks are too high, insurers exit markets."

Like insurance companies, state and local governments also are bearing the economic impacts of climate change like never before. In California alone, overtime costs for firefighters soared by 65 percent to more than $5 billion in the decade between 2008 and 2018, according to a *Los Angeles Times* investigation.[53] In 2018 alone, the state paid more than $1.3 billion in overtime to firefighters to battle record blazes. But some of the biggest costs of California's wildfires have been to the state's utilities, and, in turn, eventually to their customers who ultimately pay the bills. One month after Merced Property and Casualty's bankruptcy filing came another bankruptcy, the biggest in corporate history.

In its Chapter 11 declaration in January 2019, PG&E, the state's biggest utility, said it needed the court's protection as it figured out how to deal with an estimated $30 billion in wildfire liabilities stemming from the Camp Fire and multiple other fires sparked in part by faulty power lines stressed and stretched beyond their capacity by high winds, dry conditions, and other impacts of climate change.[54] The bankruptcy engulfed PG&E and reverberated throughout the financial and utilities sectors in new and previously unthinkable ways, ensnaring big bond holders, pension funds, and other investors. It wasn't the first time PG&E was facing billions of losses from wildfire. It had previously paid out $11 billion to insurance companies and other claimants from fires in 2017 and 2018. It paid another $1 billion to a group of fourteen cities, counties, and other public entities to offset taxpayer losses from fires.

PG&E's bankruptcy is a textbook example of how our country's failure to invest in our electricity supply system and take action on climate change resulted in economic issues of epic proportions. In the aftermath of the Camp Fire, investigators traced the origins of the blaze back to a broken century-old metal hook that cost about $1 that attached transmission lines to a tower PG&E had not inspected for nearly two decades. Yet if not for the impacts of climate change, including tinder-box conditions created by drought and extreme winds and warmer winters that increased forest-killing pine bark beetle outbreaks, the spark from PG&E's equipment failure could've been much less destructive—and much less costly.

When it comes to utilities, the costs from climate-driven wildfires and an outdated power system and grid aren't just limited to the destruction that fires leave in their wake. Increasingly, they're taking a toll on the economy even before fires begin. In mid-October 2019, as hot Diablo winds roared across the northern part of the state, PG&E made the unprecedented decision to preemptively cut the power flowing across its lines to prevent sparking—and prevent more potential liabilities. The decision left tens of thousands of homeowners and businesses in the dark, resulting in an estimated $2.5 billion in economic losses just from the blackouts because companies couldn't open, parents and children had to stay home from work and school, and financial transactions had to be delayed. In the southern part of the state, other utilities would follow suit, causing more economic losses and prompting comparisons of America's richest state to a developing country.[55]

The only businesses that stayed open in some areas were those that had big generators or solar panels and battery backup systems. It's easy to imagine what $2.5 billion in investments in distributed solar-battery systems—and smart public policies to drive their adoption—could have done to keep the lights on and the economy humming instead. In the end, the utilities' rudimentary solution of de-energizing powerlines wasn't enough to stop disaster in the face of harrowing winds and ultra-dry conditions born from climate change. On October 23, 2019, the Kincade Fire erupted near the town of Geyserville and quickly spread over nearly seventy-eight thousand acres of Sonoma County, including some parts of the region that had been burned out just a year earlier. A day later, the Tick Fire ignited north of Los Angeles and quickly roared its way across more than 4,500 acres around the nation's second-biggest city, prompting the massive evacuation of forty thousand residents, many in the middle of the night and without electricity. Among them was NBA star LeBron James, a recent Los Angeles–area transplant. As he and his family fled flames that were coming at them like a runaway fast break, basketball's King James paused for social media. "Man, these LA fires aren't no joke," he tweeted.[56]

If the fires of 2018 and 2019 seemed like something from a horror story, the wildfires of 2020 and 2021 seemed like something from the apocalypse.

California state officials knew early on that 2020 would be a rough one for the state. An extremely dry end of 2019 continued into the new year. What had traditionally been the state's rainy months, January and February, yielded very little rain or snow across much of the parched state. By March, even before the first wildfire of the year, fears of a particularly bad fire season ahead prompted Gov. Gavin Newsom to preeminently declare a state of emergency. Coupled with the emergence of COVID-19, which required state officials to rethink everything they knew about fire evacuation and shelter plans to how to staff fire departments ravaged by sickness, Newsom and his team knew that the state was headed for trouble once the weather warmed up.

They had no idea just how bad it would get.

What has become known as fire season in California typically begins in June or July, when summer ushers in hot weather and thunderstorms bring lighting. It typically lasts through November, when winds shift from blowing west-to-east and bring hot-dry east-to-west gusts from the Mojave Desert and Great Basin, often blowing at hurricane speeds. In 2020, the fire season started in May, when a two-thousand-acre blaze flared up along Interstate 5 in rural Kings County, a predominantly agricultural community east of the Sierras in the central part of the state. Within three months, more than 650 out-of-control wildfires were ranging across the state. Many had been sparked by the incredible siege of dry lightning strikes in Northern California on the morning of August 16, when an estimated 2,500 bolts ripped through the skies—including two hundred in just thirty minutes at one point—setting dry grass and forestlands ablaze as if with a box of burning matches launched one by one into the wind. As armies of firefighters simultaneously battled COVID-19 and the infernos, smoke clouded the skies and tinged the air a continent away in Washington, D.C. In satellite videos, the plume clearly was visible, streaming across the country like a ship afire on the ocean. Even if they dared leave their homes in the middle of the COVID-19 pandemic, many residents of San Francisco, Los Angeles, and countless smaller communities couldn't venture outside because the smoke was so bad. The skies above San Francisco turned an eerie blood red, prompting preachers and other evangelists to claim it was the end of days. Coupled with the pandemic, political and racial strife, and the never-before-seen weather conditions, their fire-and-brimstone prognostications were more believable than ever. By the end of the 2020 wildfire season, more than 4.3 million acres had gone up in smoke—approximately 4 percent of all the land in the entire state.[57]

Long after the embers finally cooled and the smoke cleared, the costs keep mounting, and they'll be paid not just by Californians but by everyone in every state who pays taxes, has property insurance, or buys goods produced in California—meaning, just about every American. The National Interagency Fire Center, which tracks how much the U.S. Forest Service and other federal agencies spend on fighting fires, estimates taxpayers spent about $3.6 billion nationwide on fire suppression in 2020. About $2.3 billion of that was in California alone. For perspective, the United States spent less than one-fourth that amount, about $810 million nationally, just a decade earlier. On average, the federal government spent about $1.8 billion a year on fire suppression between 2009 and 2019, about half of what it spent in 2020. Those are just the costs for federal firefighters. It doesn't include state and local firefighters and the billions more in costs to state and local taxpayers.[58]

Even with the relentless work of firefighters, 11,600 homes and other structures were lost in California alone in 2020, according to the interagency fire center. Billions of dollars in real estate were turned to ash and rubble, bankrupting businesses and families just as the country was in one of the worst economic downturns in recent history, thanks to COVID-19. But the costs extend far beyond just those homeowners and business owners who lost everything in the fires. More than fifty property insurance companies filed rate hike requests in California in 2020, seeking to raise rates on property owners by as much as 40 percent in some areas.[59] Those that didn't request rate increases tried to drop customers as soon as they could. Between 2018 and 2019, the number of property insurance nonrenewals in California soared by more than 30 percent, primarily in areas that faced the highest fire risks. After the 2020 fire season, insurance companies once again created such a stampede trying to get out of certain parts of the state that California Insurance Commissioner

Ricardo Lara had to implement a moratorium on nonrenewals, threatening to pull the license of any insurance company to do business anywhere in the nation's biggest state if they dropped too many customers. Those customers who were dropped before the moratorium took effect were forced into the state's FAIR Plan, the state-run insurer of last resort. But they weren't exempt from skyrocketing rates either: On January 1, 2021, the FAIR Plan increased rates by a statewide average of about 16 percent. That was on top of a 20 percent increase two years earlier. "In a blink, my insurance went from $1,500 per year to over $4,800, more than triple what I was paying previously," said Colleen Cross, a retired schoolteacher in Nevada County, California, during a 2020 California Department of Insurance public hearing.[60]

The increasing costs to insure homes and commercial buildings isn't limited to California by any means. To pay for climate losses in California, North Carolina, Iowa, Florida, and other states, property insurance companies are increasing rates everywhere. According to Marsh & McLennan's Global Insurance Market Index, property insurance costs across the United States have increased every quarter beginning in 2017, including a 24 percent increase in the third quarter of 2020 alone, the biggest increase since the global insurance giant began tracking such data. Rates are expected to increase even more dramatically in the years ahead, not just because of the fires in the West but also from increased storms in the East and flooding in the middle of the country. In Florida, insurance companies had more than $700 million in losses from hurricanes and other disasters in 2020. "If we're not in a crisis, I don't know what we're in," Kyle Ulrich, president and CEO of the Florida Association of Insurance Agents, told the publication *Insurance Journal.* "Those numbers . . . just simply are not sustainable."[61]

Back in California, another year of extreme heat and lack of rain had most of the state in drought again in 2021. Despite the pandemic-battered economy, lawmakers knew they had to do something to prepare for what they expected to be yet another horrific fire season. In January 2021, Gov. Newsom included an eye-popping $1 billion for wildfire and forest management. With just about every part of the state impacted in recent years by fires, legislators hardly blinked in approving the massive spending plan. In March, the state announced it was spending $81 million for an additional 1,400 firefighters to ensure there were enough reinforcements as the weather turned warmer and drier. In April, Newsom signed a $536-million wildfire package designed to help "harden" vulnerable fire-prone communities and homes by building fire breaks around them, better managing the forests surrounding them, and subsidizing upgrades for homes and other buildings to make them safer from fires. In other words, taxpayers across the state would help pay for protecting the homes of people who lived in fire-prone areas. By January 2022, Newsom was proposing an additional $1.2 billion in spending to hire more firefighters and equipment. "Climate change is making the hots hotter and the dries drier, leaving us with world record-breaking temperatures and devastating wildfires threatening our communities," Newsom said.[62] Just as the hots were getting hotter and the dries drier, the costs were getting costlier.

Costs to taxpayers aside, California utilities know they also have to do more after getting hammered by lawsuits, bankruptcies, and lost revenues in the wake of California's continuous conflagrations. So at the beginning of 2021, California's biggest utility, PG&E, announced it would spend a whopping $15 billion over the next two years to bury and upgrade their transmission lines in an attempt to try and keep them from sparking more wildfires in the future. The state's two other big utilities, Southern California Edison and Sempra Energy, announced similar plans. By the end of the year, all of the utilities were increasing rates to customers in part to to pay for the climate-related costs of burying and upgrading their power lines. The bills are just beginning: PG&E has estimated it could cost as much as $13.5 billion to put more of its power lines underground—more costs that will be passed along to ratepayers.

The dizzying increases for insurance, for firefighting, for electricity, coupled with the constant fear and danger of fire, drought, and other climate-related calamities, has pushed many Californians to the point where they realize something has to change. "The people of California are strong and resilient," California Insurance Commissioner Lara said in October 2020. "But we cannot just continue business as usual."[63]

It wasn't until 2022 that Californians and others learned just how disastrously dry the American West had become. According to a February 2022 study by climate scientists in the journal *Nature Climate Change*, record drought in 2021 helped push the region to the driest point in at least 1,200 years. Studying soil moisture levels and tree rings from California to Texas and Montana, the scientists found that the megadrought plaguing the West for the previous 22 years was the worst drought since at least the 1500s. Perhaps even more ominous, the start of 2022 came with record-high heat and record-low precipitation, with no sign of relief in sight. "The worse case scenario keeps getting worse," UCLA climate hydrologist Park Williams told the *Associated Press*.

Chapter Three

The Foundation of a New Economy

While the West burned, the Midwest flooded, and the East faced down hurricane after hurricane, President Donald Trump in 2020 was governing from an alternate universe where climate science couldn't be believed and climate disasters and the costs they bore to the American economy could be ignored.

In Trump's mind, the record wildfires plaguing California and other states could be solved by raking forest floors. After the Camp Fire in 2018, Trump traveled to California to hold an awkward press conference amid the smoldering remnants of the town of Paradise. Before the cameras, he was flanked by outgoing Gov. Jerry Brown and incoming Gov. Gavin Newsom, both Californians noticeably uncomfortable and quiet in the early stages of what would become a tortured relationship between Trump and the leaders of the nation's biggest state. "You look at other countries where they do it differently and it's a whole different story," Trump said. "I was with the president of Finland . . . and they spent a lot of time on raking and cleaning and doing things. And they don't have any problem."[1] *Make America Rake Again* became an instant Internet meme, and a point of global ridicule for the United States. Two years later, when Trump returned to California in the middle of the 2020 fire season (and the 2020 political campaign fundraising season), he again scoffed at any connection between climate change and rampant wildfires. When Wade Crowfoot, California's Secretary of Natural Resources, suggested in a meeting with Trump and Newsom that the country needed to address climate change, not just forest management, Trump was smug and dismissive. "It'll start getting cooler—you just watch," the president replied. Crowfoot said he wished the science agreed with the president's assessment, to which Trump retorted, "Well, I don't think science knows, actually."[2]

If Trump's assessment of climate change and climate-related disasters bordered on bizarre, his assessment of clean energy was simply out of touch.

Solar, he said in a speech in Iowa in 2019, was "not strong enough." Wind turbines "kill all your birds" while the noise from them somehow "causes cancer," he claimed.[3] In an appearance on the radio show of supporter Herman Cain, a one-time GOP presidential candidate who later died from COVID-19 after attending one of Trump's political rallies, Trump suggested renewable energy did not work at all, and wasn't worth it. "Unfortunately, it's not working on a large-scale," he said. "It's just not working. Solar is very, very expensive. Wind is very, very expensive."

In fact, the opposite was true. In 2019, solar and wind became the cheapest source of energy across America, with prices dropping to $40 per megawatt hour for utility-scale solar and $41 for wind, compared with $56 per megawatt hour for energy from natural gas–fired power plants, and $109 for coal-fired plants.[4] And while solar and wind prices continue to decline—thanks to improvements in technology, production scale increases, and the fact that sunshine and wind are abundant and free—the cost of electricity from coal and gas are expected to continue to rise.

When it came to energy-sector jobs, Trump seemed to live in the past. At a 2018 speech in West Virginia, he declared "the coal industry is back," even though coal jobs would fall to an all-time low during his presidency, dropping 17 percent between the time he took office and the time he left.[5] In his final State of the Union speech in February 2020, he declared energy jobs were at an all-time high, falsely suggesting it was only because of the growth in oil and gas his administration had created. He failed to even mention clean energy in the speech, despite the fact that the number of clean-energy jobs created in 2019 (70,800 jobs) was more than three times greater than the number of fossil fuel jobs created that year (21,900 jobs).[6] Nor did the president mention that clean energy accounted for more than 55 percent of the entire energy sector's job growth, or that solar and wind jobs were the two fastest-growing occupations in the whole country, according to Trump's own Bureau of Labor Statistics.[7]

If Trump and his cabinet, heavy with former fossil fuel industry lobbyists, couldn't kill clean energy with rhetoric, they tried to kill it with policy. Even as they sought to open up seemingly every inch of America to drilling and mining, from oceans to national parks, the Trump administration simultaneously tried to block expansion of clean energy at every turn. During Trump's four years in office, the Bureau of Land Management put more than 26 million acres of public lands up for lease for oil and gas drilling—some for as little as $1.50 per acre—as well as 78 million acres of offshore waters, according to research from the Center for Western Priorities and the Center for American Progress.[8] By comparison, there were fewer than 158,000 acres of public lands under lease to solar and wind projects at the end of Trump's term, BLM data shows, almost all of it because of his predecessor. Under the Obama administration, twenty-nine solar and wind projects were permitted on public lands. The Trump administration permitted a total of three solar and wind projects, not including projects that were already in process when Trump took office. The about-face on renewable energy leases and the tremendous ramp-up of oil and gas leases wasn't surprising given the backgrounds of those running the Interior Department and advising Trump. Before becoming Trump's Interior Secretary, David Bernhardt was an oil and gas industry lobbyist whose clients included Halliburton and the Independent Petroleum Association of America.[9] Acting Bureau of Land Management Director William Perry Pendley was a conservative activist who during the Reagan administration oversaw coal and gas leases under Interior Secretary James Watt, until Pendley was reassigned for vastly underpricing coal mining leases in Wyoming's Powder River Basin.[10] Scott Angelle, Trump's director of safety and environmental enforcement at Interior, made nearly $1 million serving on the board of an oil and gas pipeline company in Louisiana before joining the administration.[11] Assistant Interior Secretary Doug Domenech was also an oil and gas lobbyist who directed the Koch brothers–funded Texas Policy Foundation before joining Team Trump.[12] Scott Cameron, who oversaw Interior's policy, management, and budget operations, was previously a lobbyist whose clients included Royal Dutch Shell, the Marcellus Shale Coalition, and other petroleum interests.[13]

But as Trump and his appointees tried to kill clean energy at every opportunity during the four years he was in office, a funny thing happened: Clean energy grew faster than it had ever grown before, attracted more investments than ever, and became one of the fastest job creators in modern history. While Trump was trying to pump up coal, oil, and gas, the smart money during his time in office was fleeing to clean energy. Investments in renewable energy rose by 28 percent in 2020 alone, hitting $55 billion and making the United States the second-biggest country for clean-energy investments in the world.[14] That same year, eight coal companies filed for bankruptcy protection, including Ohio-based Murray Energy, one of the nation's largest and whose CEO Robert Murray was one of Donald Trump's biggest supporters.

Many of the investments in clean energy in recent years were driven by renewable energy tax credits signed into law by the two Republican presidents who preceded Trump and who saw clean energy as a way to improve the economy and national security, not a wedge issue to divide the public, as Trump and more recent entrants to the GOP saw it. President George H. W. Bush signed into law the 1992 Energy Policy Act that for the first time included the production tax credit (PTC) for wind and bioenergy companies. With the PTC, wind and bioenergy companies can earn a tax credit of a couple cents for every kilowatt hour of electricity they produce. It's a drop in the bucket compared with the tax breaks petroleum companies had gotten for nearly a century before the 1992 Energy Policy Act passed, but an incentive that was nonetheless instrumental to the growth of the wind energy industry, including in Bush's adopted home state of Texas, which would become the nation's biggest state for wind energy. Thirteen years later, his son President George W. Bush would sign into law an update to the Energy Policy Act that included a 30 percent investment tax credit (ITC) for solar. The younger Bush's ITC would help level the energy playing field for solar just as the elder Bush's PTC did for wind energy. Still, for more than a decade, investing in renewable energy tax credits was a lucrative cottage industry for entrepreneurs and investors—something for energy geeks and deep-pocketed environmental do-gooders who were investing in the future, not the present.

But as the price of wind and solar steadily decreased—in large part due to investments in clean-energy research and development sparked by the 2009 American Recovery and Reinvestment Act and other federal and state clean-energy policies—the interest from mainstream investors steadily increased. In early 2021, the main street bank with more branches on American street corners than just about any other, Wells Fargo, announced it had become one of the biggest investors in tax-equity clean-energy projects in the country. During the previous decade, Wells Fargo put $10 billion in more than five hundred tax-equity–financed wind, solar, and fuel cell projects in thirty-two states, representing 12 percent of all renewable energy projects in the country.[15] Like its retail banking reach, Wells Fargo's investments went far beyond the clean-energy hotbeds of California and New York. One of the biggest projects it bankrolled in 2020 was a 227-megawatt solar project on the banks of the Tennessee River in Muscle Shoals, Alabama, that would become that state's biggest solar project. Wells Fargo estimated the project would create three hundred jobs, generate $1 million in local taxes, and supply enough power to keep the lights on in hundreds of thousands of homes. The banking company also invested $350 million into a 2.6-gigawatt solar project in Spotsylvania, Virginia, not far from where the nation's first coal mine opened in 1709. The Spotsylvania Energy Center project, expected to be the biggest solar project east of the Rockies, was built to provide electricity to Microsoft Corp.'s data centers and other operations as part of the computing giant's commitment to get 100 percent of its energy from renewable sources. Locally, the project also generated about seven hundred local construction jobs and millions in local property tax dollars.

Few companies illustrate the economic transformation of the electricity industry and the rush of investment dollars to clean energy better than the aptly named NextEra Energy Inc. Based in Juno Beach, Florida, the company got started as Florida Power & Light in the 1920s, riding the coattails and railroads of Standard Oil Co. co-founder Henry Flagler, whose turn-of-the-century investments and boosterism transformed southeast Florida from a mosquito-infested swampland into one of the most populous parts of the country. For most of its history, Florida Power & Light was a small local company, but one not afraid to tackle big ideas. While its primary business in its early years was bringing the still relatively new innovation of electricity to the growing number of homes being built from Miami to Palm Beach for northern transplants who rode Flagler's railroad south in search of sunshine and warmth, in 1926, it branched out into another early venture, starting the Florida Power & Light Ice Department to deliver blocks of ice by horse and carriage to residents to keep sweltering new Floridians' homes and iceboxes cool in the days before air conditioning. The power company also opened an ice cream business to sweeten the relationship with its customers.[16]

In Florida, NextEra and its FPL subsidiary are viewed as an old-school utility that runs on fossil fuels and nuclear power. It is a pariah in the eyes of many environmentalists and regulators alike for its anti-clean-energy tactics and its environmental track record. Its Turkey Point nuclear plant near Miami has been the source of numerous environmental disasters, including radioactive leaks into once-pristine Biscayne Bay, severe algae blooms in the waters surrounding the plant, and threats to drinking water sources that supply the Florida Keys.[17] In recent years, FPL further drew the ire of clean-energy advocates and environmentalists for continuing to stifle the solar industry in the Sunshine State, including its funding and orchestration of a massively misleading ballot initiative in 2016 designed to eliminate the ability for homeowners and other owners of rooftop solar to sell excess power to the grid, a threat to FPL's monopoly.[18] Yet while NextEra and FPL have masterfully played and benefited from the regulated Florida utilities market that keeps it and the rest of the state dependent on fossil fuels, outside Florida, NextEra is a leader in clean energy. It first began investing in renewable energy in the 1990s, shortly after a bipartisan group of ninety-three of one hundred U.S. senators, including then-Senator Joe Biden, passed the 1992 Energy Policy Act that was signed by President George H. W. Bush, which in addition to creating the first tax credits for renewable energy also paved the way for the deregulation of the electricity business. Along with tax credits, NextEra (then still called FPL Energy) saw the promise and opportunity to expand outside Florida, especially after the Federal Energy Regulatory Commission began to open wholesale power markets up around the country beyond local utilities. The company began taking some of the vast amounts of the money it generated from dirty fuels in fast-growing Florida and investing it into clean wind and solar farms on the other side of the continent. In 1998, it flipped the switch on its first wind operation, the Vansycle Ridge Wind Farm near Helix, Oregon. Soon, it would pour hundreds of millions of dollars into wind operations in Texas, California, Iowa, and other states, while also investing in solar in just about every part of the country.

By 2009, when it changed its name to NextEra, the company claimed to be the world's largest producer of wind and solar energy (outside of China).[19] It also branched out into Canada. In April 2020, the company announced on a call with investors it was taking its clean-energy investments to the next level, disclosing plans to spend $1 billion on battery and energy storage projects in 2021. As wind and solar—and now, battery storage—becomes increasingly cheaper to produce, NextEra's stock has grown exponentially, and its investors have grown rich.[20] A $1,000 investment in NextEra's stock in 2005, when it began to dramatically expand its clean-energy portfolio, was worth $7,900 in 2021. By late 2021, the market value of NextEra hit $167 billion, making it more valuable than Goldman Sachs, Citigroup, or American Express.[21]

NextEra's soaring stock price is next to nothing, however, compared to the highest-flying clean-energy investment in history, Tesla Inc. Just as energy policies from Washington led to NextEra's monumental growth, Tesla's rise can be traced back to the last time the U.S. government and Joe Biden made a concerted effort to address climate change through investments in clean energy and clean transportation. The company got its start in 2003 when founders Martin Eberhard and Marc Tarpenning set about to build the country's first full-sized all-electric sedan, based not on traditional auto industry methodology but on new technology.[22] It got a major boost when fellow Silicon Valley entrepreneur Elon Musk and others injected $7.5 million in capital into the company a year later. But the company didn't really take off until after the U.S. government invested $465 million into Tesla in January 2010 through the Department of Energy's loan guarantee program. The program was created by President George W. Bush but didn't really get going until the passage of the 2009 American Recovery and Reinvestment Act that then-Vice President Joe Biden administered.[23] Six months after receiving its DOE loan (which the company would pay back in full three years later and ahead of schedule), Tesla went public at $17 per share. It was the first time an American automaker had gone public since Ford in 1956. Ten years later, Tesla's market valuation would exceed that of century-old Ford, General Motors, and Fiat Chrysler *combined,* and in October 2021, its market cap reached $1 trillion, making it one of the most valuable companies in the world. Along the way, Tesla revolutionized the American auto industry, pulling every other domestic and foreign automaker along behind it to embrace electric vehicles. In January 2021, GM announced it was planning to sell only electric and other zero-emission cars and trucks by 2035. Ford, Chrysler, Volkswagen, Mercedes-Benz and just about every other car maker is now headed in the same direction.

Despite years of false starts, the foundation for a clean-energy and clean-transportation economy were in place. It happened quicker than even the savviest captains of American capitalism had expected, and at a time when the country was in the throes of a pandemic and economic meltdown. In 2020, about 76 percent of all new energy added in the United States came from wind and solar, according to the U.S. Energy Information Administration. In 2022, wind and solar was expected to account for about 63 percent of all new electricity-generating capacity, with batteries accounting for another 11 percent of all capacity.[24]

As clean energy grew, so did the jobs that come with it. Between 2017 and 2019, clean-energy jobs grew by 6 percent, more than double the overall employment growth in the United States during that time.[25] By comparison, jobs in coal declined by more than 7 percent. By the start of the 2020s, clean energy and its three-million-plus jobs became *the* energy industry in America, accounting for nearly three times as many jobs as fossil fuels and more growth and more potential than any other part of the energy sector. And while right-leaning politicians from conservative states sometimes try to push the false narrative that clean-energy jobs are something for left-leaning liberal states like California, beginning in 2020, Texas was the no. 2 state for clean-energy jobs (behind California), and Florida was no. 3. Rounding out the top ten states for clean-energy jobs are other red states such as Ohio, North Carolina, and Virginia. Pennsylvania, the birthplace of the oil industry in America, is home to about ninety thousand clean-energy jobs, making it the no. 11 state

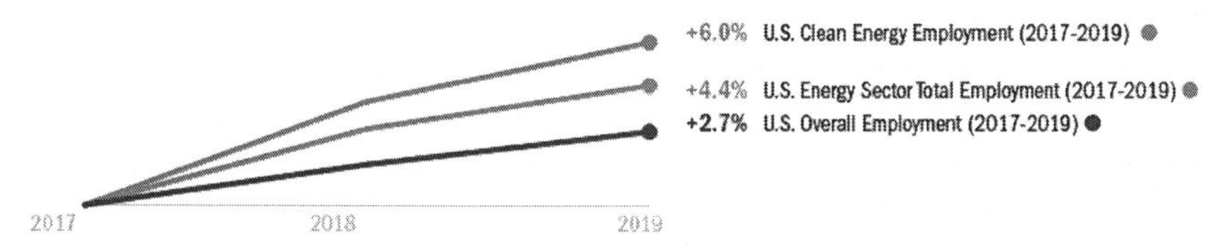

Figure 3.1. Job-growth rates by energy sector, 2017–2019. *Source*: E2, ACORE, CELI, BW Research, "Clean Jobs, Better Jobs (2021)."

for clean-energy employment, followed closely by Indiana, the conservative home state of Trump's vice president, Mike Pence.

Even in states known for fossil fuels, clean-energy jobs now outnumber coal, oil, and gas jobs, whether or not their elected representatives acknowledge it. Senator Mitch McConnell has long been one of the biggest defenders and proponents of coal in Congress, understandably so given the rich and important history of coal in his home state of Kentucky. For years, one of the centerpieces in the main meeting room of McConnell's office in the Russell Senate Office Building was the bust of a coal miner that looks a little like him, even though the career politician has never worked a day in coal himself. He isn't alone. The fact is, fewer people now work in coal mining in McConnell's home state of Kentucky than work in clean energy and clean transportation —by a factor of seven. In 2020, about 4,500 Kentuckians worked in coal mining, according to the Kentucky Office of Energy Policy, compared with more than 33,000 who work in renewable energy, energy efficiency, grid and battery storage, and clean vehicles.[26] Among the 1,500 or so solar industry workers in Kentucky are those employed by Bluegrass Solar, which in 2017 was hired to install an eighty-panel solar array on top of the Kentucky Coal Museum in Benham.[27] "The irony is pretty prevalent," Bluegrass Solar owner Tre Sexton told EKB-TV. Financially, it's also pretty smart. The move to clean energy was expected to save the coal museum and the surrounding community as much as $10,000 a year in energy costs. Clean cars also are creating jobs in Kentucky. In 2020, twice as many Kentuckians produced hybrid cars than produced coal in the state, thanks in part to Toyota's eight-thousand-employee Georgetown factory that makes hybrid RAV4 and Lexus ES cars. In January 2021, amid the COVID-19 pandemic and one of the worst job markets in recent history, Hitachi Automotive Electric Motor Systems America announced it was opening an electric motor factory in Berea, Kentucky, creating two hundred new, local jobs.[28] In September 2021, Ford and partner SK Innovation announced they would invest nearly $6 billion in central Kentucky to build the BlueOval SK Battery Park.[29] There, more than five thousand workers are expected to crank out batteries and other electric vehicle parts to power Ford's lineup of electric vehicles. Ford also announced it was investing nearly $6 billion in a mega-campus in Stanton, Tennessee. where it will build electric F-150s, and another $90 million for a factory in Texas.

Kentucky, Texas, Tennessee, and other states illustrate that, unlike coal or petroleum jobs, clean-energy jobs aren't constrained by geology, geography, or politics. Despite the demonization of clean energy by many in the Republican Party, there were almost as many clean-energy jobs in Republican congressional districts than there were in Democratic congressional districts in 2021.[30] An analysis by E2 showed that 54 percent of all clean-energy jobs were in Democratic districts while 45 percent were in Republican districts. And while the nation's largest metropolitan areas are home to most clean-energy jobs, among the metro areas with the largest share of clean-energy jobs in their workforce, twenty of the top twenty-five have populations of less than five hundred thousand. Among the mid-sized metros leading the country in clean-energy jobs are Holland–Grand Haven, Michigan, and Cleveland, Tennessee, both of which are buoyed by thousands of jobs connected to the electric vehicle and energy efficiency industries and both of which elected hard-right, conservative Republicans in 2020 to represent them in Congress. Perhaps the biggest clean-energy story is little Storey County, Nevada. There, clean energy supports more than two jobs for every employable resident, making it the county with the highest density of clean-energy jobs in the country. More than eight thousand people worked in Storey County as of early 2020 in what has become the hottest sector in clean energy: Batteries and storage. Tesla started the battery boom when it broke ground on its massive Gigafactory battery plant on the aptly named Electric Avenue in 2016, and went on to hire seven thousand employees there.[31] Other companies followed, including battery recycling company LiNiCo and data center operator Switch, which in July 2020 broke ground on one of the largest solar and battery storage projects in the country in Storey County.[32] The project, the crown jewel of Switch CEO Rob Roy's "Gigawatt Nevada" vision, will include thousands of solar panels and buildings filled with Tesla's batteries, allowing the project to generate 555 megawatts of solar energy when the sun is shining—enough juice to power more than ninety-one thousand homes—and store 800 megawatt hours that can be deployed after the sun goes down. Like other EV and clean-energy hotbeds, Storey County in recent years has been represented by a Republican, U.S. Rep. Mark Amodei, the former chairman of the Nevada Republican Party who was first elected in 2011.

While jobs in solar, wind, and battery technology may be the first that come to mind when clean-energy occupations are mentioned, the industry's reach goes far beyond solar and wind. The biggest sector of clean energy is the one that's usually most overlooked: Energy efficiency. In 2021, about 2.4 million Americans worked in energy efficiency related jobs.[33] They include sheet metal workers, pipefitters, and heating, ventilation, and air conditioning (HVAC) technicians who install high-efficiency heating and cooling systems in homes, offices, and schools across America. They're electricians who design and install high-efficiency LED lighting systems and programmable thermostats and temperature sensors that can dramatically reduce the electricity costs in skyscrapers in Manhattan and homes in Minnesota. They're construction and insulation company workers who install Low-E windows and high-density insulation, and plumbers who replace inefficient hot water

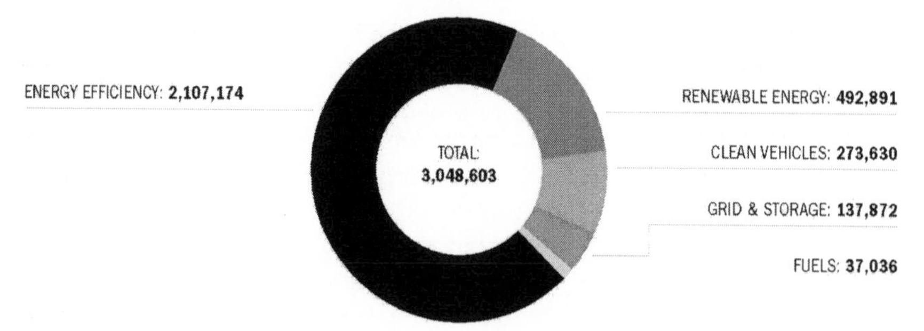

Figure 3.2. U.S. clean-energy employment, by sector. *Source*: E2, "Clean Jobs America, 2021."

heaters with high-efficiency tankless and heat pump models. They're factory workers who make building efficiency products and Energy Star appliances, such as refrigerators and dishwashers, which today use a fraction of the electricity they did a decade ago, or commercial combined heat and power units that enable factories or fossil-fuel power plants to capture wasted heat generated by heavy machinery and turn it into steam or hot water.

They're people who work at places like Energy Optimizers USA in Dayton, Ohio. The company was founded in 2009, just as the country was coming out of the Great Recession and as the Obama administration was just beginning to push clean energy as a solution to get Americans back to work and to fight climate change. Energy Optimizers specializes in improving energy efficiency in public schools and colleges. Its workers and contractors swap out inefficient lighting with energy-sipping LED lighting, replace old boilers with high-efficiency HVAC systems, install better windows and insulation, and hook it all up to solar panels where it makes sense. The savings can be significant. At the Monroe Local Schools District in Middletown, Ohio, for instance, Energy Optimizers workers and contractors installed LED lighting with automation and controls that turn off the heat and air when nobody is around and regulate the electricity used by the walk-in freezers and other appliances in cafeteria kitchens. All simple stuff. But as a result, Monroe Local Schools has saved an estimated $108,000 per year in annual energy expenses, about an 18 percent return on the costs it spent for the upgrades, just in the first five years. "The investment in energy-saving improvements just made smart fiscal sense," Superintendent Dr. Phil Cagwin told Energy Optimizers.[34] "We've improved our energy efficiency, air quality, and building comfort, all with no cost to our taxpayers." Along with the Monroe Local Schools project, Energy Optimizers tracks the energy and environmental impacts on just about every project it undertakes. In its first decade in business, the company estimates it saved school districts in Ohio and surrounding states more than $100 million, enough to help create or preserve the jobs of 1,500 teachers. It also estimates it reduced carbon emissions by 2.1 billion pounds, the equivalent of planting 382,000 acres of trees.

Don't think for a minute that Energy Optimizers founder Greg Smith is some hippy tree hugger. A giant of a man whose hobbies include bear hunting and riding dirt bikes, Smith got his start selling air conditioning and heating systems for Trane and other manufacturers across Ohio. He's quick to point out he's a right-leaning independent voter and a fiscal conservative. And he shies away from calling the jobs he created at Energy Optimizers anything but what they really are. "Don't call them clean-energy workers," said Smith, who has since sold the company. "These are electricians, pipefitters, HVAC techs, sheet metal workers," in addition to engineers, architects, and computer programmers. In other words, they're traditional blue-collar, made-in-America jobs that just so happen to be in energy efficiency occupations.

Call them what you want, these workers are saving Americans money with every project. Energy Star appliances alone save American consumers and businesses nearly $40 billion in annual energy costs, an amount equivalent to the entire gross state product of Vermont.[35] LED lighting, which uses about 85 percent less energy than incandescent bulbs, is expected to save another $30 billion by 2027, according to U.S. Department of Energy estimates. In the typical American home, installing low-E storm windows can cut annual power bills by up to 33 percent, saving the average household almost $450 a year. Double the attic insulation in your house, and you'll save another $1,800 over ten years.

If energy efficiency is the workhorse and the biggest part of the clean-energy sector, solar and wind are the stars. At the beginning of 2020, before the COVID-19 pandemic struck, nearly 523,000 Americans worked in renewable energy, including solar (345,000 workers) and wind (115,000). Solar and wind are naturally what politicians and the everyday public think about when they think of clean energy. A photo op standing next to rows of shining solar panels or beneath towering wind turbines is simply a lot sexier than being down in the basement with an HVAC technician swapping out an old boiler for a high-efficiency, combined heat and power unit or hanging out in the attic with an insulation installer.

That's what President Obama and his handlers had in mind in May 2009 when he landed at Nellis Air Force Base in Nevada for a tour of what was then the biggest solar array in the country. With more than seventy thousand solar panels spread over 140 acres, the original project supplied about 25 percent of the energy used by the base and the twelve thousand military and civilian

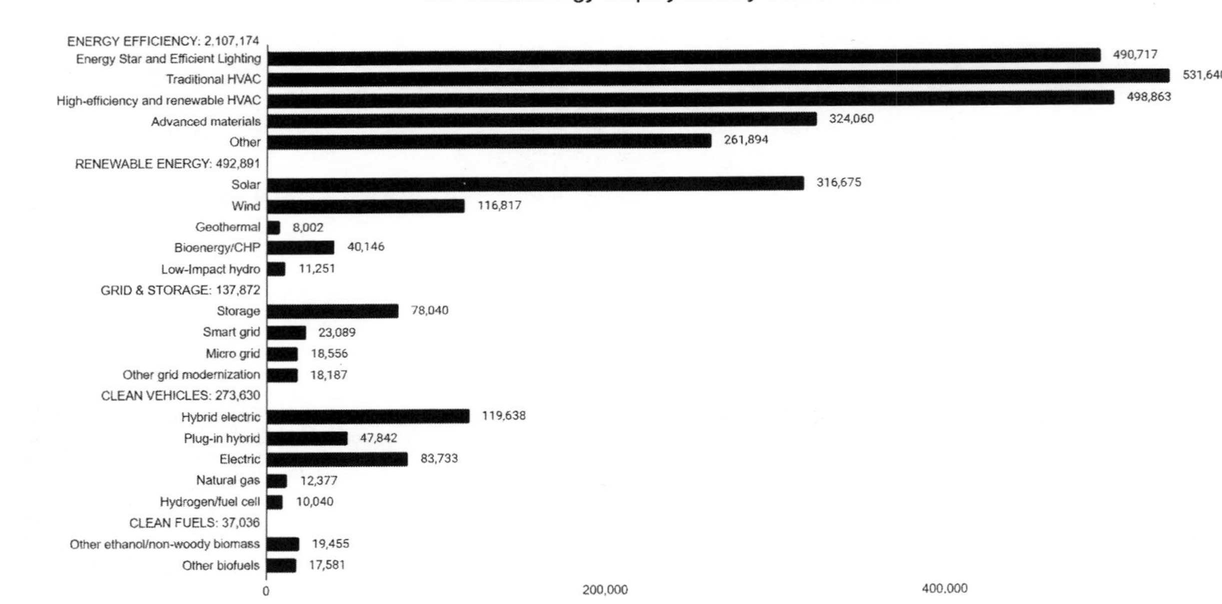

Figure 3.3. U.S. clean-energy employment, by subsector. *Source*: E2, "Clean Jobs America 2021."

personnel there. It was later doubled in size, making Nellis completely energy independent on sunny days. The project actually got its start in 2007—the year before Obama was elected—but there wasn't a better backdrop in the country to show what Obama's American Recovery and Reinvestment Act could do to spur clean-energy investment and growth, create jobs, reduce our dependence on fossil fuels, and increase our national security. "We showed how we can accomplish the goals of strengthening the military, creating jobs, and keeping our country safe," says retired U.S. Air Force Col. Dave Belote, who as commander of the 99th Air Base Wing at Nellis at the time was in charge of showing Obama and Nevada Senator Harry Reid around shortly after the base flipped the switch on the solar project. Belote is a proud Virginian who grew up in Virginia Beach, spending summers swimming in the tributaries of the Elizabeth and James rivers, despite the pollution from nearby creosote plants that populated the region back then. After graduating from the University of Virginia (or as Belote reverentially refers to it, "Mr. Jefferson's University"), he joined the Air Force, where he flew F-16s. "I spent most of my time as a young pilot trying to turn petroleum into jet noise as fast as I possibly could," he says. Experiencing first-hand what pollution and sea level rise did to his native state and seeing the potential of solar during his time at Nellis, Belote became a firm believer in the importance of clean energy for the country's future.

After leaving Nellis, he went to the Pentagon to set up the Department of Defense's Siting Clearinghouse that reviews all applications for renewable projects near military bases. Later, he went to work for a major wind energy company before starting his own clean-energy and government affairs consulting firm and investing in offshore wind. He also happens to be a five-time *Jeopardy!* champion. For Belote, the connections between clean energy and national security are clear, and they have nothing to do with politics. "That this has become a partisan issue boggles my mind," he says. "It's simple: Burning fossil fuels drives destabilizing climate change, which will put more of our men and women in uniform in harm's way."

Like Belote, Troy Van Beek got religion about clean energy while in the military. Tall, steely-eyed, and Hollywood-handsome, Van Beek spent nine years in the Navy SEAL teams. Most of that time, the teams were in the Persian Gulf, protecting oil tankers in the Strait of Hormuz. It didn't take long for him to realize the massive amounts of money and the thousands of lives being risked every day guarding those shipments. He later started a private security firm whose business included providing protection to former Afghan president Hamid Karzai. Death and destruction eventually rattled him. Seeking peace, he ended up in 2006 at Maharishi University of Management near Des Moines, Iowa. There, he met wife Amy, the daughter of a local homebuilder.

Wanting to apply their conscience-based business ideals to the real world, the couple in 2009 started Ideal Energy, an energy efficiency company they later transformed into a solar company just as the Obama-Biden American Recovery and Reinvestment Act was rolling out. "I started realizing I could have a different impact on world security—with energy," Van Beek says. Like Belote and like legions of other military veterans, Van Beek sees the direct connection between clean energy and national security. "My time in the military gave me a perspective on why we go to war," he says. "And since then, I've been really impassioned to eliminate the need to go to war. A lot of people think that's impossible, but I've come to believe it's actually possible as we move to an energy abundance paradigm for the world (with clean energy). War is just a matter of scarcity. When we eliminate scarcity, we eliminate the need for war."

Ariel Fan came into the clean-energy business from a much different place.[36] A passionate environmentalist since her youth—at age thirteen, she helped lead a Greenpeace campaign to stop logging in rainforests—one of her first paying jobs was doing energy audits and giving away free LED lightbulbs to New York City small businesses as part of a Con Edison energy efficiency program. Day after day, she would walk from bagel shop to dry cleaner to corner grocery store with clipboard in hand, trying to convince busy business owners they could save a lot of money on their electricity bill if only they took the time to fill out some paperwork and swap out some lightbulbs. The experience helped Fan realize doing good for the environment could also be good business. After graduating from Columbia University, she returned home to California, where she launched a project to make Brighton Management's sixty-plus hotels more energy efficient and sustainable by converting to LED lighting, low-flow toilets, and closely managed recycling programs. In 2017, at the age of twenty-four, Fan started her own company, GreenWealth Energy, with the idea of expanding her energy efficiency work beyond the hospitality industry. In September 2020, however, she saw the promise in something else. California Gov. Gavin Newsom signed an executive order designed to phase out gas-powered vehicles by requiring all new cars and trucks sold in the state to be electric or otherwise zero-emission vehicles by 2035. The policy and the way the auto industry was going prompted Fan to fully shift her business model to one of the fastest-growing clean-energy sectors: Electric vehicle charging.

Today, Fan's Los Angeles–based GreenWealth Energy helps businesses, local governments, and other organizations deploy electric vehicle charging stations and switch to all-electric fleets by doing everything from applying for grants and tax credits to doing the construction and installation to later managing the units for its customers. Her company is the first certified women and minority-owned (Fan is Taiwanese-American) electric vehicle charging company in California, and one of her goals is to bring more women and minorities into clean energy. In its first two years in the car-charging business, GreenWealth installed more than five hundred charging stations throughout California to help businesses and other organizations meet state requirements while also generating new income sources at their properties. By 2022, encouraged by plans by both federal and state governments to create a nationwide car-charging network, Fan was raising capital to expand outside her home state, while also trying to keep up with growing demand within California.

Fan's work is a small part of one of the biggest shifts in the American economy in history. "We're at the precipice of the complete overhaul of a one-hundred-plus-year-old industry—the auto industry—that has relied on fossil fuels and combustion engines forever," she said. "This is not just a once-in-a-generation event but a once-in-a-multi-generation event, where we completely transform the way we think about transportation and vehicles and the waste and impact we have on our planet."

While clean energy is one of the fastest-growing and most promising job sectors in America, Ariel Fan is an anomaly. Clean energy has a diversity problem.[37] By far, it is an industry dominated by white men. Only about 26 percent of clean-energy workers are women, even though women represent about half of the U.S. population. Asians represent about 7 percent of the clean-energy workforce, about equal to the number of Asians in the workforce overall. Hispanics/Latinos represent about 18 percent of the overall

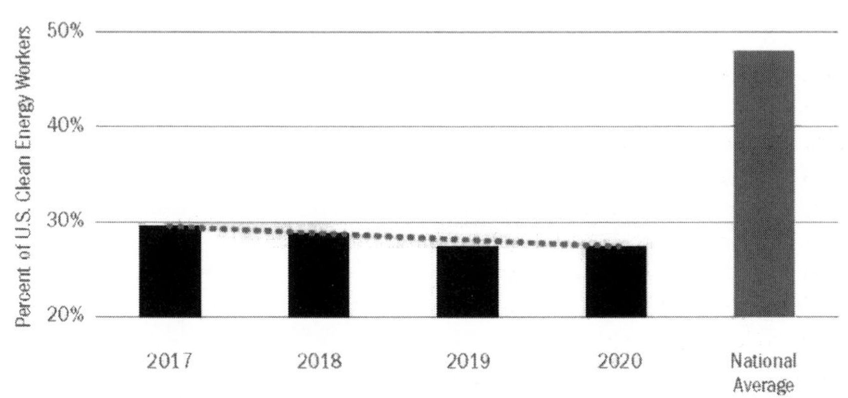

Figure 3.4. Women in clean energy. ***Source:*** **E2, AABE, BOSS, Alliance to Save Energy, Energy Efficiency for All, BW Research, "Help Wanted: Diversity in Clean Energy (2021)."**

U.S. workforce but only about 16 percent of the clean-energy workforce. Still, that's better than being Black and in clean energy. Only 8 percent of clean-energy workers are Black, compared with about 12 percent of the overall workforce. Overall, more than 75 percent of all clean-energy workers (and all workers in energy fields including oil, gas, and nuclear) are white.

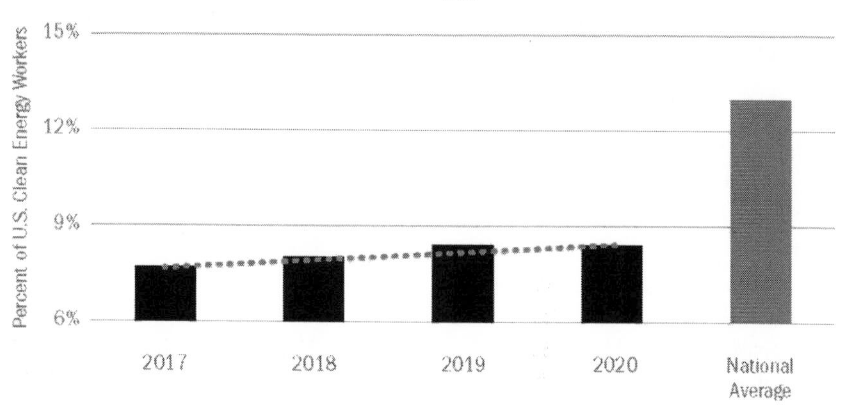

Figure 3.5. Blacks in clean energy. *Source*: E2, AABE, BOSS, Alliance to Save Energy, Energy Efficiency for All, BW Research, "Help Wanted: Diversity in Clean Energy (2021)."

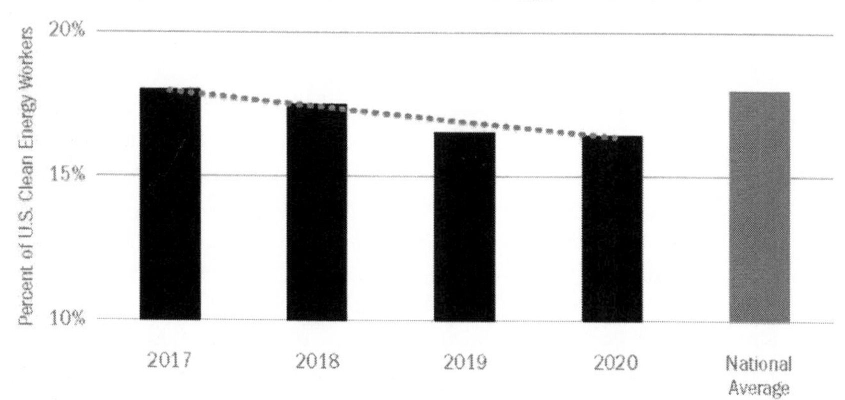

Figure 3.6. Hispanics and Latinos in clean energy. *Source*: E2, AABE, BOSS, Alliance to Save Energy, Energy Efficiency for All, BW Research, "Help Wanted: Diversity in Clean Energy (2021)."

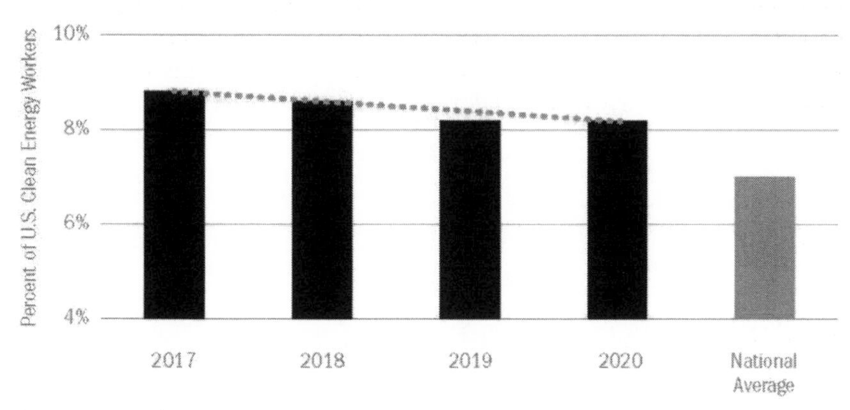

Figure 3.7. Asians in clean energy. *Source*: E2, AABE, BOSS, Alliance to Save Energy, Energy Efficiency for All, BW Research, "Help Wanted: Diversity in Clean Energy (2021)."

Paula Glover spent fifteen years in the utilities industry before she became president of the nonprofit American Association of Blacks in Energy, and later, the Alliance to Save Energy. She said that while clean-energy jobs are becoming more common for people of color, more needs to be done. Blacks in clean energy, she said, are typically part of what she calls the "onlys"—the only person of color in a room filled with white people or the only woman in a room of men. But with people of color and young people now representing an increasing amount of the American population overall, clean-energy companies need to change their hiring practices too if they want to survive, she said.

"That is what the future looks like," Glover said. "So as a sector, if you are planning to be there in another ten, fifteen, twenty years, and you're not planning to make your workforce more diverse, I am not sure what you do, because the students coming in are more diverse."

Racial inequity, the need for new jobs after the COVID-related economic meltdown, national security, appealing to conservatives, creating unity—these were all issues bouncing around in Joe Biden's mind in the spring of 2019 as he launched his fourth attempt to become president of the United States. For him, it quickly became clear that expanding clean energy and addressing climate change were a big part of the solutions to many of these problems. But even more importantly, as he approached his seventy-eighth birthday after four decades of public service, Biden also knew in his mind and in every bone in his body that addressing climate change and expanding clean energy were simply the right things to do. Global warming had been a challenge Biden knew America needed to confront ever since he first learned about the issue and its impacts on America's economy, its national security, and its competitiveness beginning in his earliest days in the Senate. Later, as vice president and administrator of the American Recovery and Reinvestment Act, he saw first-hand how government support of clean energy could create jobs and drive industry-changing innovation.

For Biden, the immediate problem in the heat of the 2020 presidential campaign was that others now knew these things too, including the growing number of other Democratic politicians coming out of the woodwork to challenge Donald Trump for the presidency, and who were all well ahead of Biden when it came to talking about the impacts of climate change and the solutions needed to address it.

Chapter Four

Reality, Intruding

The world was grappling with the explosions of the Challenger space shuttle over Florida and the Chernobyl nuclear plant in Russia. President Ronald Reagan, smarting from public backlash over his administration's attempts to open public lands to drilling and gut the Clean Air Act and Clean Water Act, had turned his attention from tearing down environmental regulations to tearing down the Berlin Wall. And in Congress during that summer of 1986, Republican Senator John Chafee convened a very unusual hearing.[1]

Chafee was a Rhode Islander, the state's favorite son, an all-American boy who went off to Yale before joining the Marines during World War II. He fought at Guadalcanal and Okinawa, commanded a rifle company during the Korean War, and ultimately served as Secretary of the Navy under Richard Nixon. Chafee also served two terms as governor of Rhode Island and in 1976 became the first Republican to win a Rhode Island Senate election since before the Great Depression. In 1984, amid the rise of Ronald Reagan, his fellow Republicans elected him chairman of the Republican Conference, making him the third-highest-ranking Republican in the Senate. But for all his political prowess, Chafee, with his thick Yankee accent and messy mop of gray hair, was known less for his politics and more for his desire for bipartisan compromise. Back before the idea of protecting the planet and its future was twisted into a partisan issue, Chafee was also one of America's most strident congressional champions for the environment.

Entitled "Ozone Depletion, the Greenhouse Gas Effect and Climate Change," the two-day hearing that Chafee gaveled to order as chairman of the Subcommittee on Environmental Pollution on June 10, 1986, promised to be yet another bureaucratic blab-fest about something most of the American public neither understood nor cared about. Most of the nation's attention that day was focused on the public release of the findings of a presidential commission on the Challenger disaster, identifying the failure of an O-ring rocket joint as the cause of the explosion. Also garnering the attention of the news media and public that day was the trial of Cathy Smith, a Canadian singer who pled guilty to involuntary manslaughter for supplying the drugs that killed iconic comedian John Belushi.

By then, the connection between rising carbon dioxide emissions and the warming planet was already clear. Three years earlier, in 1983, the Environmental Protection Agency published a report predicting the Earth's temperatures would begin rising noticeably beginning in the 1990s because of increasing carbon dioxide emissions.[2] If temperatures continued to rise unabated, EPA scientists suggested, it could have catastrophic impacts. Crazy weather was something Americans were beginning to increasingly experience. The 1980s started with a horrific drought in the Southeast that caused an estimated $20 billion in crop losses and ten thousand deaths. In 1983 came a fluke freeze in Florida that wiped out $2 billion in citrus crops. A hurricane hit Texas that same year (the first in three years to make landfall in the United States) causing $3 billion in damage. Record flooding in the West from an El Nino event caused another $1 billion in economic costs.

By the time of Chafee's hearing, the media, led by groundbreaking *New York Times* science reporter Walter Sullivan, was writing regularly about "the greenhouse effect," introducing the public to a phenomenon that climate scientists were understanding better each year. "Carbon dioxide in the atmosphere acts like the glass in a greenhouse," Sullivan wrote as far back as 1975.[3] "It readily permits sunlight to reach and warm the lower atmosphere and the earth's surface, but it impedes the escape of that heat into space as infrared radiation." Ten years later, Sullivan helped break the news of another alarming climatic issue: The discovery by British scientists of a hole in the ozone layer over Antarctica, a place that Sullivan had visited seven times as a journalist.[4] The ozone layer, Sullivan and others explained, was important because it helped shield the Earth from harmful ultraviolet rays from the sun, which on an individual level causes skin cancer and on a global level causes atmospheric circulation anomalies that lead to even more climate disasters. Coming at a time when *Star Wars* captured the public's imagination about the future and the Challenger disaster made the dangers of it more real, the media coverage of the hole in the ozone layer was raising public concern that in turn was raising concern among lawmakers like Chafee. It also was forcing scientists, the public, and politicians to begin thinking about the bigger but related issues of climate change and deforestation. And after several years of record droughts, weird freezes, and flooding from previously unheard-of atmospheric events, some were beginning to make the connections between climate change, the ozone layer, and deforestation to everyday American life. The question was, what to do about it, and when?

Chafee's June 1986 hearing, held as Washington, D.C. coincidentally broiled in a heat wave, was designed not so much to unveil new science but to finally begin to set the path for policy to address what science was telling us. "This is not a matter of Chicken Little telling us the sky is falling," he said in opening the hearing.[5] "The scientific evidence, some of which we will hear today, is telling us we have a problem, a serious problem." Among those called to speak at the hearing were NASA scientist James Hansen, Princeton University scientist Michael Oppenheimer, and the newly elected senator from Tennessee, Al Gore. If Chafee represented the power and prestige of the Senate's past, the old Rhode Island politico understood, Gore represented the future and the generation that would have to deal with the problems left behind by previous generations. When it came to global warming, Chafee also understood the connection to both the environment and the economy, and he also understood the costs and dangers of inaction. "By not making policy choices today, by sticking to a 'wait and see' approach, we may in fact be making a passive choice," he said at his 1986 hearing. "By allowing these gases to continue to build up in the atmosphere, this generation may be committing all of us to severe economic—and environmental—disruption without ever having decided that the value of 'business as usual' is worth the risks."

Joe Biden was not part of Chafee's committee or the hearing in June 1986, but Biden knew his fellow small-state senator well. From his work on the Senate Foreign Relations Committee, Biden also was beginning to understand the connections between climate, national security, and international relations better than most. At the same time the country was coming to grips with a warming world and the potentially catastrophic conditions that could result, the United States also found itself in a precarious new position in the world. In 1985, the country became a debtor nation for the first time since before World War I, the result of massive foreign borrowing by the government to pull the nation out of a recession and to pay for then-President Reagan's tax cuts. Two years earlier, a suicide bomber attacked the U.S. embassy in Beirut, killing more than three hundred military and other personnel, most of them Americans. And the nation's top adversary, the Soviet Union, was grappling with civil unrest and rising political uncertainty as it was shakily falling apart. International tensions were high, America's foreign relations were strained, and, like a sweaty heat wave in the middle of a Southern summer, the specter of global warming was increasing anxiety and discomfort about the world's future.

Three months after Chafee's seminal hearing, Biden took to the Senate floor to introduce the country's first-ever climate legislation, the 1986 Global

Climate Protection Act.[6] It received little fanfare, just another mundane bill introduced on another mundane day in Congress, wedged in between a discussion of the naming of a federal courthouse in Alexandria, Virginia, and amendments to the Social Security Act to include small rural hospitals. To try to make his legislation more relevant and urgent, Biden noted Chafee's hearing, which had been well-covered by the media and helped make climate celebrities out of scientist Hansen and new Senator Gore. Biden also noted the parallels between how future generations would have to deal with the massive foreign debt the United States was taking on and the massive climate problems the country was creating. As bad and irresponsible as it was to shackle future generations with huge national debt, he said, shackling them with the consequences of global warming was a threat to the future of mankind itself. "In an age when mankind's numbers and activities are so expensive as to affect the planetary conditions that have been the very basis of human life, the cost of irresponsibility could be survival itself," he said on the Senate floor.[7]

Biden's legislation directed the president to establish a task force to research and assess the economic, societal, and environmental threats of global warming and to develop and implement a coordinated national strategy to address those threats. The legislation called for the task force to be led by the secretary of state, underscoring Biden's appreciation of the importance of foreign relations and international cooperation to address an issue affecting not just the United States but the entire planet. It also directed the president to appoint an international climate ambassador and to make climate negotiations a part of an upcoming summit with the Soviet Union. Biden's bill went nowhere. A year later, he reintroduced it, once again referencing Chafee's hearings and once again urging action as the consequences of climate change continued to become clearer and more ominous, even in the short year since he first introduced the bill. "Even though decades away, the most serious consequences of global warming could prove unavoidable unless we act now to prevent them," Biden presciently warned his fellow senators in January 1987.[8] "Our failure to show foresight when the dangers are clearly discernible would be an unforgivable dereliction of duty to our children and all mankind not yet born." Once again, Biden's legislation died in committee.

Though unsuccessful, Biden's proposed legislation—coupled with the general growing public concern about the impacts of mankind's burning of fossil fuels—sounded the alarm for oil, gas, and coal companies and their deep-pocketed lobbyists. Biden by then was in his third Senate term. With Democrats regaining control of Congress that year, he and other powerful politicians who cared about climate change were back in power in Washington. And while Biden's initial run for president in 1984 had been a disaster—he finished a very distant sixth in the Democratic primary that year—it raised the profile and visibility of the Delaware senator. Even Republican President Reagan, who had just won reelection in a landslide, was suddenly raising concerns about ozone layer depletion and its impacts on climate and health. In 1988, Reagan joined other world leaders in signing the Montreal Protocol to ban ozone-depleting chemicals, going against advisors who tried to pretend there wasn't a problem, including then-Interior Secretary Donald Hodel, who suggested Americans just needed to wear more sunscreen and hats to avoid UV exposure.[9] The fossil fuel industry mobilized. In the years ahead, it would lodge a concerted disinformation campaign to sow seeds of doubt about climate change and tamp down the smoldering embers of climate action in Congress. It worked like a snake-charmer's spell, wooing much of the public and many lawmakers into thinking that climate change was still something in a galaxy and a future far, far away, or as conservative Republican Senator Ron Johnson of Wisconsin suggested more crudely and simply as recently as 2021, "bullshit."[10]

As a result, a little over three decades after Senator John Chafee held one of the first hearings on climate change and Senator Joe Biden introduced the first piece of legislation to address it, the predictions and warnings about inaction that the two senators, as well as a legion of scientists, environmentalists, and other forward-thinking lawmakers, made back in the 1980s came true. The year Chafee held his hearing, the United States suffered less than $6 billion in economic losses from major climate disasters; in 2021, the number was twenty-four times that amount at $145 billion, according to NOAA

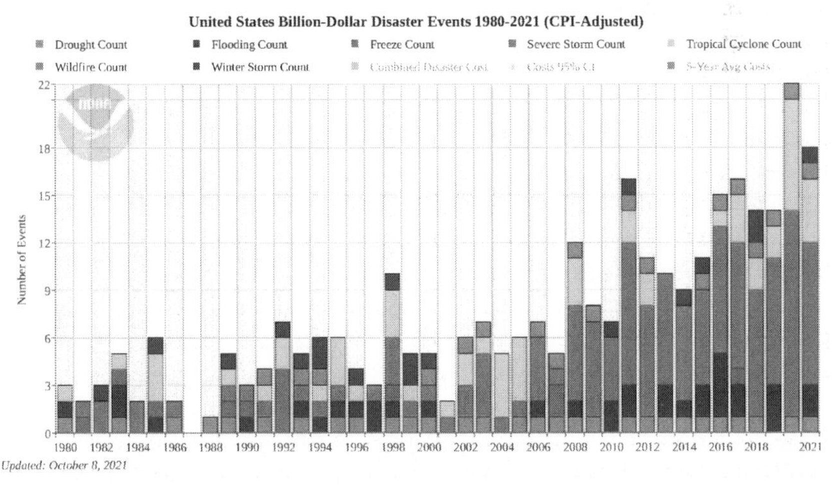

Figure 4.1. U.S. billion-dollar climate-related disasters annually, by type and number. *Source*: NOAA National Centers for Environmental Information (NCEI), "U.S. Billion-Dollar Weather and Climate Disasters (2021)."

statistics. In the 2020s, there is no doubt that record heat, record wildfires, record storms, record ice melt, record flooding, and other growing calamities are the product of climate change and the nation's ambivalence toward addressing it. It isn't "bullshit" as Senator Johnson deemed; it is "a serious problem" as Senator Chafee suggested in 1986. "Reality has a way of intruding," Biden would later say, referencing his Global Climate Protection Act from three decades earlier.[11]

By the start of the 2020s there was no doubt the reality of climate change was intruding on something else: the economy. Record financial losses from weather-related disasters and record declines in profits by oil, gas, and coal companies, coupled with record investments in renewable energy and electric vehicles, were changing the narrative. Climate change was now an economic issue. And when something becomes an economic issue, it also becomes a political issue. Throughout the history of U.S. presidential elections, climate change barely registered in importance with the public or with candidates. In 2008, the economy and the federal budget deficit ranked as the top issues for voters, followed by terrorism, energy, and Iraq, polling showed.[12] In 2012, the issues remained almost the same. Climate change didn't even register among voters surveyed by most polling organizations in either election year. In 2016, the economy as usual led the top priorities for voters, followed by terrorism, foreign policy, and health care. In 2020, however, climate change burst into the presidential political race like never before. In a January 2020 poll by Gallup, 55 percent of respondents ranked climate change as one of the most important issues in the election, ranking higher than traditional hot-button issues like abortion, taxes, immigration, and foreign relations.[13] In a poll by Pew five months later, two-thirds of respondents said climate change was affecting their local community and that the federal government was doing too little about it.[14] By October 2020, one month before the election and in the aftermath of a horrendous year of fires, floods, drought, and hurricanes, fully 60 percent of adults surveyed by Pew said they viewed climate change as a major threat to the well-being of the United States, up from 44 percent just a year earlier.[15] Climate change was the no. 4 issue among those surveyed, preceded (in order) by the economy, health care, and the coronavirus pandemic.

Yet what polls didn't measure was how climate change had also become a big part of the no. 1 issue consistently in voters' minds—the economy. By 2020, voters in every part of America had seen, experienced, or knew someone who suffered financial losses from climate-related disasters. If it wasn't a hurricane wiping out houses and businesses in North Carolina, it was wildfires burning them down in Oregon. If it wasn't flooding in Iowa ruining farms and car dealerships, it was tornadoes blowing them away in Tennessee. Yet Americans were also beginning to see something else for themselves: The economic benefits of climate action. By 2020, most Americans had seen workers installing solar panels on a rooftop somewhere, had driven by wind turbines in the open plains, or had been passed by electric vehicles on the open road. Consumers and business owners alike saw the savings on their own monthly power bills from installing LED light bulbs or Energy Star appliances. If seeing was believing, hearing was reinforcing. And suddenly, Americans were hearing from nearly every candidate for president, with the notable exception of incumbent Donald Trump, that addressing climate change and expanding clean energy was critical to getting Americans back to work and getting an economy deflated by COVID pumped up again. For decades, disinformation campaigns by the fossil fuel industry, the U.S. Chamber of Commerce, and conservative groups and politicians had spoon-fed many Americans the false choice that they had to pick between a good environment or a good economy. But after experiencing the economic costs of climate change themselves, after seeing or receiving the economic benefits of clean energy, after hearing the promise of a cleaner, brighter economy from politicians with every political ad, Americans in 2020 were finally seeing past the fossil fuel smokescreen. A poll later taken by CBS News found that about 75 percent of Americans said they thought the climate and the economy were linked. A plurality, 45 percent, said that increasing efforts to stop climate change would help the economy, compared with 30 percent who said it might hurt the economy.[16]

On the Democratic campaign trail, it was as if the 2020 presidential contest was all about the candidates trying to outdo each other on climate action. For an issue that had barely registered with voters in the past, it was suddenly front and center. Washington governor and former Congressman Jay Inslee was first out of the blocks, announcing in May 2019 a $9-trillion climate action plan that included massive national public transit projects, a GI Bill–style retraining program for displaced coal workers, and a promise to create eight million new clean-energy jobs through his "Evergreen Economic Plan." Senator Bernie Sanders said he would declare a "national climate emergency" and invest a whopping $16 trillion into climate and clean-energy solutions and the jobs that came with them. Senator Elizabeth Warren rolled out a plan of plans that included everything from a "Green Apollo Program" to a "Green Marshall Plan," invoking imagery of the massive build-up of the aerospace industry during the space race or the rebuilding of Europe after World War II. Former South Bend, Indiana, mayor Pete Buttigieg, burnishing his experience as a business consultant, threw in the big idea of a climate bank to fund clean energy. Long-shot candidate Tom Steyer, a billionaire California businessman, spoke of almost nothing else but climate action and creating clean-energy jobs during his brief time on the campaign trail. And eventual vice president Kamala Harris, a former prosecutor, said she would help pay for her $10-trillion climate action plan by prosecuting fossil fuel companies for economic and environmental damages.

Despite his early leadership on climate in the Senate, Joe Biden was the slowest of any Democratic candidate to include climate action in his platform during the 2020 presidential election. The old Washington war horse knew climate change never polled well in the past and he didn't see the shift that had occurred. So instead, he chose to make his campaign about his experience and the sharp contrast between him and President Trump, especially on matters of humility and morality. After three years of self-inflicted mistakes and fumbles by a president and an administration staffed with sycophants who had little or no experience in Washington—much less running the country—mainstream America was tiring of Trump's on-the-job training and demagoguery. They were ready for somebody with experience who knew what they were doing and who wouldn't embarrass the country on the international stage. Biden rode his message of experience and morality to the top of the polls for months, but it wasn't enough to keep him there. While he was clearly winning with over-fifty voters, he was losing with everybody else. In June 2019, a Monmouth University poll found that Biden's favorability with Democratic voters under fifty had fallen to a mere 6 percent.[17] It wasn't just about his age: Bernie Sanders was the candidate most liked by under-fifty voters, and Sanders was a year older than him. Biden knew he had to do something differently, and knew he had to step up his game on climate change. If he didn't get the message from the polls, he got it loud and clear in December 2019, when the Sunrise Movement of young people for climate action passed out "Green New Deal" report cards for the Democratic front-runners. Sanders earned an A- from the group. Warren got a B−. Biden got an F−. As Sunrise put it, it was "parent-teacher conference time" for Biden.[18]

To help him figure out how to appeal to younger voters, Biden turned to a younger voter—one he trusted more than most. Stefanie "Stef" Feldman had been an intern for then-Vice President Biden in 2010 when she was still a senior at Duke University's Sanford School of Public Policy. Despite their differences—she was a twenty-something from Smyrna, Georgia, and he was an old man from Delaware—the two hit it off, quickly building a relationship akin to a proud grandfather and doting granddaughter. After graduating from Duke, Feldman was offered a job in the vice president's office and soon became his deputy director for economic policy. Barely out of college, Feldman suddenly found herself briefing the vice president of the United States regularly on policy issues ranging from the environment to gun control to veterans services. After the election of 2016, she had planned to go to law school at Yale. But a few months before she was to start, she got a call from

Biden asking her to come to another university—the University of Delaware. Biden was launching his Biden Institute at the university's school of public policy (also named after him), and he wanted Feldman to be its director. Two years later, when Biden shelved his plans to ride off into the political sunset and instead got back on the campaign trail, Feldman joined him as national policy director for the Biden-Harris campaign.

Feldman and other young people on the campaign didn't need an F- rating from Sunrise Movement to make it clear to them that Biden needed to focus more on climate change if he wanted to attract younger voters. Sanders had been riding a wave of popularity among younger voters since the 2016 election, and—while free college was always appealing—it was largely because younger people saw in him a politician who finally understood the need for climate action. Feldman understood it too: She and her fellow millennials, after all, would be the ones inheriting the mess left behind by earlier generations. But while Sanders wanted to take a socialist sledgehammer to the foundations of the American economy to address climate change, Feldman wanted to find another way that would be more appealing not just to young voters but to a broader swath of the population. Growing up in the red-state South, she was particularly interested in figuring out how to pull up the parts of America left behind by technology and a changing economy along with the rest of the country.

While at Duke, she spent a summer in eastern Kentucky, where she wrote her senior thesis about the prospects for economic development in rural Appalachia. Among other places, she studied Pike County, the biggest and easternmost county in Kentucky, which was—like many other parts of Appalachia—dependent on the coal industry. While other coal counties withered away with the decline of the industry, Feldman noted in her thesis, Pike County fared fairly well.[19] It became a hub for the state's banking industry and attracted a Kellogg's cereal factory as well as a budding telecommunications firm, creating new jobs and new careers for coal miners and their families. What helped make Pike County different, Feldman noted, was political leadership. Instead of just complaining about the loss of coal jobs, Pike County leaders went out and recruited new industries and businesses, and then created policy to help those businesses relocate and thrive. And not surprisingly, once the businesses and jobs came, the politicians who brought them won the most votes in local elections.

At Biden's direction, she began formulating campaign strategy that would broach the intersection of environment and economy. After years of working for Biden, she understood what he wanted in a policy. "The thing I learned from him the most was when I briefed him for the first time many years ago," Feldman told C-SPAN in December 2020.[20] "I told him about this policy we were going to do, and he said 'That's great. How will that actually look on the ground (to) real people? How are you actually delivering results for people?'"

"I didn't have that great of an answer for him then, and I never made that mistake again," Feldman told C-SPAN. "Now, we spend a lot of time iterating back and forth about how we talk about (policies) in ways that make sense to Americans, and how we make sure we're actually delivering."

Biden and Feldman knew that when it came to climate change, they didn't want to talk about parts per million or carbon taxes or even about specific technologies, like solar or wind. Unlike Sanders, they didn't want to demonize the fossil fuel industry—there were too many potential swing Republicans and right-leaning Democrats in places like Texas and Colorado and the South who they'd alienate with talk like that. Unlike Warren, they knew they didn't want a complex plan of plans. Climate change was complicated enough. Why make it harder?

By the summer of 2020, with the economy in the tank thanks to COVID-19 and the economic costs of climate disaster after disaster making it the worst, most expensive year for climate-related damage the country had ever seen, the message became clear for Biden and Feldman. Biden had started his latest campaign for the presidency promising to save the soul of the country. Now, it looked like he also needed to save the economy of the country as well as the future of the planet. After months of tiptoeing around the subject of climate change, Biden settled on making clean energy and climate action the centerpiece of his campaign—and if he were lucky enough, the centerpiece of his presidency.

On July 14, he took the stage in Wilmington, Delaware, for what aides signaled would be his most important speech of the campaign.[21] But the signals were mixed, and at least initially, confusing for some members of the press. Would Biden be speaking about his economic recovery plans, or his climate plans? Biden would make the answer clear in his remarks as he went on to detail a $2-trillion economic recovery plan built on climate action.

"When I think of climate change," he said, "the word I think of is jobs."

It turned out to be a winning message.

Chapter Five

Beyond Business as Usual

Politicians weren't the only ones making the link between climate change and the economy in 2020 in ways that will ultimately impact consumers, businesses, and, eventually, the planet.

Since 2012, BlackRock chairman and CEO Larry Fink has written an annual letter to fellow CEOs. His missives are part reflection, part prediction, and all about business vision and values. As the head of the world's biggest investment firm, with nearly $9 trillion in global assets under management as of 2021, the release of Fink's letter each January is akin to a Manhattan Moses's delivery of the stone tablets, if they happened to come in an annual edition. For most of his decade or so of doing this, Fink has focused on esoteric business and management issues: The value of long-term results versus short-term returns, how tax law changes fit into investment strategy, what roles companies play in their communities, or how to improve corporate governance. But in 2020, Fink's letter was different. Titled "A Fundamental Reshaping of Finance," it had little to do with balance sheets or cost-benefit analyses and everything to do with droughts, hurricanes, wildfires, and flooding.[1]

"Climate risk is investment risk," wrote Fink, a mantra he would later echo in media interviews, speaking engagements, and elsewhere. Fundamentally, Fink explained, the nation's financial foundation was being shaken to its core by climate change. And if CEOs and investors didn't do more to address it, Fink predicted, dire economic times lay ahead. "Over the 40 years of my career in finance, I have witnessed a number of financial crises and challenges—the inflation spikes of the 1970s and early 1980s, the Asian currency crisis in 1997, the dot-com bubble, and the global financial crisis," he wrote. "Even when these episodes lasted for many years, they were all, in the broad scheme of things, short-term in nature."

"Climate change is different," Fink wrote. "Even if only a fraction of the projected impacts is realized, this is a much more structural, long-term crisis. Companies, investors, and governments must prepare for a significant reallocation of capital."

The message resonated with businesses around the globe.

A few weeks after Fink's letter, Delta Airlines said it would invest $1 billion to go carbon neutral by 2030.[2] Salesforce CEO Marc Benioff teamed up with renowned environmentalist Jane Goodall to launch a program to plant a trillion trees.[3] And in perhaps the boldest announcement of any corporation, Microsoft announced it would not only cut its carbon emissions to zero, it would do better by going carbon negative by 2030, meaning it plans to remove more carbon from the atmosphere than it produces by then.[4] By 2050, the software giant pledged, it would go a step further, removing from the atmosphere as much carbon as it had emitted since the company was founded in 1975.

Microsoft's announcement in particular was met with a healthy dose of skepticism in the media and beyond, although not so much so among those who knew the company and the festering concerns of its founder and other top officers about climate change. Microsoft, in fact, had been a leader in corporate climate responsibility for some time. More than a decade earlier, the company promoted a passionate young program manager named Tamara "TJ" DiCaprio to help oversee its fledgling sustainability operations. DiCaprio had majored in environmental studies and geography at the University of California, Santa Barbara. She worked a few years as a commodities trader at Merrill Lynch before joining Microsoft in 2005 to do corporate marketing and sales.[5] She had relatively little formal training in sustainability—something that wasn't unusual back then—but she did have a big idea, which at a place like Microsoft was as important as a degree from any university. By then, Microsoft's value to the economy was already as big as many states and a few countries. Its nearly ninety thousand employees were spread across 175 office locations in ninety countries, including the fifteen million square feet of office space at its five-hundred-acre headquarters in Redmond, Washington. With such a large carbon footprint, Microsoft could surely make a difference in the world if it took some steps to reduce its emissions.

At the time, California had just passed its landmark 2006 Global Warming Solutions Act, which required the state to adopt a "cap-and-trade" system that included caps on greenhouse gas emissions and set up a system for big polluters to either reduce their emissions or pay for credits, which in turn could be used by the state to invest in clean energy and other emissions-reduction efforts. Following California's lead, other states were beginning to consider similar legislation, and two climate champions in the U.S. Congress, Rep. Henry Waxman of California and Rep. Ed Markey of Massachusetts, were drafting legislation for a federal cap-and-trade program modeled after the California law.

DiCaprio's big idea was to do something similar with businesses, starting with Microsoft. Under the idea she posited to her bosses, Microsoft would set up its own carbon-emissions caps on a rolling basis.[6] In turn, each division at Microsoft would have its own emissions caps they had to stay below. If a division exceeded its cap, it would have to pay what essentially amounted to an internal fine to the company, which in turn could be used to pay for sustainability efforts across the company or buy carbon offsets to reduce its overall corporate carbon footprint. The way the company managed its internal carbon program was relatively simple and not dissimilar to the way it handled the distribution of other company assets, like, say, printer paper. Each year, every Microsoft division received an allocation of carbon credits, just like it received a budget allocation for printer paper. If divisions exceed their allocation, division managers would have to cut into other parts of their budgets to pay for more of them, just like they would have to buy more printer paper if they exceeded their annual allocation for paper. If a division exceeded its annual carbon cap too many times, the division manager would be held accountable as part of their annual performance review. Of course, enforcing such a program with senior company managers would take leadership from the top, but that wasn't a problem: DiCaprio's internal carbon-fee idea was just the kind of metrics-based program that Microsoft managers loved.

The carbon-fee program was an early cornerstone in the foundation for the broader strategy Microsoft president Brad Smith announced in January 2020 on the heels of Fink's "A Fundamental Reshaping of Finance" letter. In addition to continuing to set more stringent carbon caps each year, Smith announced, Microsoft would take other major steps to reduce its emissions

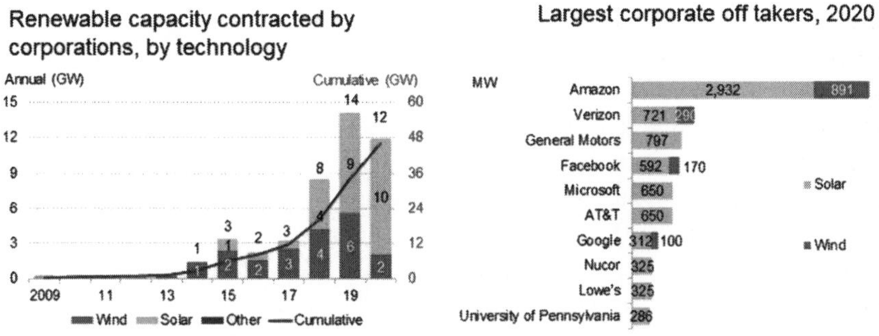

Figure 5.1. Large companies are some of the biggest buyers of clean energy in America. *Source*: Bloomberg NEF, the Business Council for Sustainable Energy (BCSE), *Sustainable Energy in America 2021 Factbook*.

across every "scope" of its operations. In climate speak, "Scope 1" emissions are generally considered direct emissions from a company and the sources it controls, such as the emissions generated from heating and cooling its offices and factories, operating its vehicles, or running its equipment. "Scope 2" emissions come from the generation of electricity or fuel it purchases, while "Scope 3" emissions generally refer to those a company produces indirectly from everything from business travel on airplanes and in cars to the carbon produced by the meat or other foods it serves in its company cafeterias to the waste it generates from doing business. Scope 3 emissions reductions are always the hardest to tackle.

To reduce its Scope 1 and Scope 2 emissions, Smith announced, Microsoft would power its facilities and operations with 100 percent renewable energy by 2025 and electrify its entire fleet of vehicles by 2030. To tackle the tougher issue of reducing Scope 3 emissions, Microsoft turned back to the program DiCaprio started. Smith announced that beginning in July 2020, the company's internal carbon-fee program would be expanded so that every division would not only have to stay below caps on their direct emissions and pay an internal fee if they exceeded them but also do the same for their indirect Scope 3 emissions, including everything from the jet fuel burned for employee travel to emissions generated by suppliers of everything from microprocessors to plastic game controllers. If that wasn't hard enough, Smith's new directive took it a step further: Division managers would now also have to account for the electricity use over the life cycle of the products they produced, meaning every Microsoft Xbox and every piece of software in every computer that ran Windows and every data center with a Microsoft-powered server. In doing so, the program would have an added benefit: It would encourage managers to make more energy efficient products, which in turn could help reduce the emissions (and the electricity costs) for anyone who uses a Microsoft product.

Fink's 2020 letter to CEOs focused on the economic costs of climate change, but his January 2021 letter focused on the other side of the coin, the economic benefits of climate action. Shortly after President Biden took office, Fink wrote about the acceleration of "a tectonic shift" of money to more sustainable assets, including clean energy, something he admitted was happening "even faster than I had anticipated." "We know that climate risk is investment risk," he reminded clients, customers, and fellow CEOs in January 2021. "But we also believe the climate transition presents a historic investment opportunity."

Once again, companies and corporate titans reacted. Amazon earmarked $2 billion to invest in clean-energy and clean-transportation projects, including a major infusion into electric vehicle maker Rivian to build the replacements for the online retailer's near-ubiquitous delivery vans.[7] Jeff Bezos, Amazon's founder, donated another nearly $800 million to sixteen environmental and climate change organizations. GM CEO Mary Barra, following the lead of other carmakers and taking heed of the signals from the incoming Biden administration, pledged to convert all of the automaker's production to electric or other net-zero vehicles by 2035, pointing out that doing so would also create and protect thousands of good-paying union jobs.[8] Microsoft cofounder Bill Gates wrote a book about the transformational power of clean-energy innovation. By the end of 2021, more than 2,100 corporations were working with the Science Based Targets Initiative (SBTI), a nonprofit organization designed to help companies set greenhouse gas emissions goals in line with the Paris Agreement designed to curb global warming by well below 2 degrees Celsius.[9] About half of those companies had already set science-based emissions targets, according to SBTI. Setting such goals makes simple business sense. Companies that set goals based on science-based targets—meaning they do more than just pay lip service and set arbitrary emissions-reduction goals—save money, boost investor confidence, spur innovation and competitiveness, and keep regulators at bay from imposing new laws that could impact their business. The number of companies setting such goals is growing quickly: According to SBTI, from November 2019 to October 2020, about thirty-one companies joined the initiative every month on average, more than double the rate between 2015 and 2019.

Like BlackRock's Fink and Microsoft's Smith, other titans of Wall Street and industry also stepped up their climate game. When Jane Fraser became CEO of banking and investment giant Citigroup Inc. on March 1, 2021, her first communication to the public wasn't about interest rates or lending forecasts or even about her landmark achievement becoming the first woman CEO of a major Wall Street bank. It was about climate change.[10] On her first day on the job, Fraser announced Citigroup's commitment to have net-zero greenhouse gas emissions by 2050 by powering all its facilities with renewables and reducing its Scope 2 and Scope 3 emissions. Nine months later, she also warned that to meet its goals, the financial giant may also drop some of its clients, hinting that it may quit doing business altogether with oil and gas companies and other big carbon emitters. "At the end of the day that will mean there are some choices as to which clients we will be serving and which ones we won't be," she told the *Wall Street Journal*.[11] "One-size-fits-all won't work."

Finally, corporate America seemed to be getting the message that climate change was an economic issue. The warnings and predictions from scientists and environmentalists that businesses had ignored for decades were now ringing loudly in the ears of CEOs all across the country. But this time it came not from activists or environmentalists but from the highest prophets of profits. The message was not wrapped around a tree or illustrated with photos of polar bears but swathed in the sanctity of the almighty dollar. If they weren't hearing it from business leaders like Fink and Fraser and Gates and Benioff and Bezos, company leaders were hearing that they needed to do more from investors, their managers, and their employees.

Beginning in 2020, investors pulled money out of fossil fuel companies in record amounts. Simultaneously they invested it into environmental, social, and governance (ESG) funds in record amounts. In 2020 and again 2021, investments in U.S. exchange-traded funds that focus on clean-energy sustainability reached new highs. Investors sank more than $330 billion into sustainable funds through September 2021—nearly double the amount invested in such funds a year earlier, according to financial research firm Morningstar Inc.[12] A record thirty-eight new sustainability funds launched in the third quarter of 2021 as investor interest rose along with federal action and the growing realization of the economic costs of climate change and the economic benefits of climate action.

Still, financial firms around the world are sitting on trillions of assets invested in oil, gas, coal, and other carbon-polluting companies, putting their own financial future and that of their shareholders and customers at risk. According to Moody's Investors Service, financial firms at the end of 2021 held about $22 trillion in loans and investments subject to carbon risk—or, as *Bloomberg News* called it in November 2021, a "$22 Trillion Carbon Time Bomb."[13] Moody's advised that banks and other financial institutions that don't shift their lending away from fossil fuel companies and into clean-energy and climate-focused companies could face major risks to their credit quality in the future.[14] If that happens, banks won't be able access funding as cheaply and, in turn, will have to make more loans at higher rates, hurting their profitability, their business models, and their shareholders.

Already, shareholders are increasingly demanding that companies do more to reduce their carbon emissions and address climate change. According to the business advocacy group Ceres, more than 135 climate-related shareholder resolutions were filed in the first quarter of 2021 in advance of annual meetings at companies in industries ranging from pizza to petroleum, all pushing them to do more to improve sustainability and climate action.[15]

Billionaire British investor Christopher Hohn took matters into his own hands. Unhappy with how big companies were addressing climate change, he began using his TCI Fund Management Ltd., one of the most profitable hedge funds in the world, to pressure companies it held interests in to develop, publish, and meet science-based carbon-reduction plans. Later, he and his charitable foundation, the Children's Investment Fund Foundation, started a broader shareholder campaign called “Say on Climate,” and teamed up with California-based nonprofit shareholder advocacy group As You Sow to use shareholder resolutions to pressure at least one hundred of the S&P 500 companies to adopt science-based initiatives to reduce carbon emissions.[16] In one big example of the campaign’s effectiveness, more than 90 percent of the shareholders of food services and distribution company Sysco Corp. in late November 2021 voted for a resolution demanding the company to create a plan within one year to disclose and reduce greenhouse gas emissions targets to get the company to net-zero emissions as soon as possible. It wasn’t as if Sysco was ignoring climate change and the impacts on its business. Earlier that year, the company announced plans to electrify 35 percent of its delivery fleet—the equivalent of 2,500 electric trucks—and reduce its overall emissions by more than 27 percent by 2030. For Hohn and As You Sow, however, that wasn’t nearly enough. The company needed to set and implement company-wide greenhouse gas emissions and reductions goals in line with the Paris Agreement. “Piecemeal efforts to reduce emissions are insufficient to achieve our global climate goals,” said David Shugar, ESG and climate data analyst at As You Sow.[17] “Shareholders recognize that climate-related financial risk to companies and to shareholder portfolios continues to grow. Inaction or limited action is no longer acceptable.”

By the end of 2021, the “Say on Climate” initiative funded by Hohn and managed domestically by As You Sow had engaged with seventy-five of the nation’s biggest companies, with more on the horizon. “You only need to buy $25,000 of stock and hold it for one year to file a shareholder resolution in the U.S. and $2,000 in Canada,” Hohn told the *Wall Street Journal*. “With $12 million, you can buy enough shares to file them with every company in the S&P 500.”

Bill Weihl took a different approach. An MIT-trained computer scientist, Weihl cut his teeth in the technology business as a research intern for IBM before moving on to engineering jobs at Digital Equipment Corp. and later, Akamai. In the early 2000s, he began to take a keen interest in climate change, reading everything he could about it, getting involved with organizations such as the Sierra Club and E2 and spending a lot of time thinking about it from his technologist perspective. “I became increasingly convinced that this could be a major crisis,” Weihl said. “And I decided I wanted to devote not just my personal life but my professional life to address it.”

In 2005, Weihl left his job as the chief technology officer of Akamai, a cutting-edge cloud computing and cybersecurity company, and one year later joined upstart Google as its green energy czar. Google at the time was building data centers around the world and beginning to grapple with the double-edged problem of getting enough cheap electricity to operate those centers while also limiting the environmental impacts of using so much energy. Weihl launched the company's initiative to get 100 percent of its energy from clean, renewable sources. But he also realized back then there wasn't enough renewable energy available at a cheap enough price. Under Weihl's leadership Google helped change that, too. It began funding external and internal research and development projects to help drive rapid innovation in renewables, with the goal of making it cheaper than coal as quickly as possible. He and Google also figured out new ways of contracting for cleaner energy, using virtual power purchase agreements with utilities where it had operations.

In 2012, Weihl moved to Facebook, where he oversaw similar efforts as the company's director of sustainability. By then, solar and wind was getting cheaper, but some states and their utilities were still dragging their feet on shifting away from coal and other fossil fuels. So Weihl decided to set a new rule: In order for Facebook to consider any state for a data center, it had to have an ample supply of renewable energy. And if there wasn't enough renewable energy available, Weihl would tell local politicians who were itching for a ribbon-cutting ceremony that they'd have to pass some regulations to encourage utilities to produce more renewable energy. The strategy helped change the power industry regulations in many states—places like North Carolina, Texas, Alabama, and Georgia—where clean energy was previously considered just something for greenies out in California. Weihl also extended his efforts beyond his own companies, working with the World Wildlife Fund and other NGOs in 2013 to help start the Renewable Energy Buyers Alliance (REBA), an organization that today includes more than two hundred major technology and other companies committed to powering their operations with renewable energy and using their collective clout to push economic development officials and lawmakers to make more renewable energy available in their states.

In 2015, Weihl had an epiphany while watching another social issue play out that had nothing to do with clean energy or climate. In Indiana that year, then-Gov. Mike Pence signed the Indiana Religious Freedom Restoration Act (RFRA) that made it legal for companies and individuals to discriminate against lesbian, gay, bisexual, and transgender people based on religious beliefs.[18] The law was a thinly veiled pushback to the gay marriage movement that was gaining momentum nationwide. Around the same time, the state of North Carolina passed its infamous "bathroom bill" that declared people in the state could only use single-gender bathrooms that corresponded with the sex listed on their birth certificates. In other words, transgender people or those who self-identified as something different than what was listed on their birth certificate couldn't use the bathroom they wanted to use. In addition to garnering immense criticism from gay rights and gay marriage activists nationwide, the Indiana and North Carolina laws resulted in a huge backlash from businesses, sports teams, and other organizations. PayPal, Deutsche Bank, and other tech and banking businesses cancelled plans to open new offices in North Carolina, taking billions of dollars in investments and thousands hundreds of jobs with them.[19] Salesforce canceled plans to expand a major operation in Indiana. Numerous sporting events and concerns were called off or moved to other states, sucking millions out of the states' economies.

Weihl saw all of this and thought about climate change. If companies could use their collective voice to demand change on gay rights and same-sex marriage, he thought, they could use their voice to demand climate action. In February 2020 Weihl founded ClimateVoice, a nonprofit that seeks to organize and empower workers at technology and other companies nationally and help them push their companies to step up, speak out, and get active in demanding climate policy changes at the state and federal level. Weihl wants workers at major companies across America to force management to go all-in on climate, both by improving their internal sustainability practices and by engaging in policy advocacy in Washington, D.C., and in statehouses across the country. ClimateVoice's first projects included clean-energy legislation in Virginia and Illinois and clean transportation legislation in New York and other Northeastern states. In 2021, it launched a new campaign to encourage tech workers and customers of Alphabet (Google's parent company), Amazon, Apple, Facebook, and Microsoft to sign petitions urging the companies to devote at least one out of five of their lobbying dollars toward lobbying for federal legislation to keep global warming below 1.5 degrees Celsius. It also launched an online scorecard that lets activists see if companies support the Biden administration's climate and clean energy initiatives and use social media and emails to pressure the company to do so if not.[20]

"Policy is about influence, and businesses have a lot of influence," Weihl said. "If businesses say they want good carbon policies, we'll get them."

Increasingly, businesses say they want good carbon policies.

Shortly after Biden took office in January 2021, about 450 business leaders from across the country signed a letter from E2 urging Congress to set aside partisanship and ideology and work with the new administration to prioritize clean-energy jobs and investments in climate action—or else continue to watch America lose ground to China and other countries on clean-energy innovation, investments, and jobs.[21] The signers included individual investors, CEOs, and small-business leaders from every region of the country. Another group of more than four hundred businesses ranging from Starbucks and McDonald's to Ford and Coca-Cola signed a letter drafted by Ceres urging

Congress to pass legislation that would reduce U.S. greenhouse gas emissions by at least 50 percent by 2030, in keeping with the new Biden administration's goals.[22]

Businesses now want action on climate change not only because they see the economic benefits of shifting to a cleaner economy but also because they're clearly scared about what will happen if we don't. Each year, the World Economic Forum surveys experts and leaders in insurance, finance, and other businesses, along with academics, government leaders, and NGOs, to gauge their assessment of the biggest risks to the world. A decade or so ago, the issue of climate change typically ranked low in the Forum's annual Global Risks Report, but that's changed in recent years. In 2021, respondents said the no. 2 risk facing the world was "climate action failure."[23] Only "infectious diseases" ranked higher in the list of global worries in 2021, which was not surprising during the middle of a global pandemic. "Climate action failure" ranked higher than "weapons of mass destruction" and more than thirty other world threats. Respondents also ranked other climate issues including extreme weather, biodiversity loss, and natural resources crises among the biggest risks to the world, well above issues like terrorist attacks, the collapse of Social Security, and commodity and price instability. And unlike with the COVID-19 crisis, the World Economic Forum noted in its 2021 Global Risks Report, there's "no vaccine for environmental degradation."

By the time the Biden administration was settling into the White House, businesses were lined up, politicians were lined up and the world was on fire. The economics of climate change were hitting home. Yet the most seasoned members of Biden's team—and there were many—knew that big ideas and optimism only go so far in Washington. Along with environmental advocates and all but the newest members of Congress, they knew from experience that even the most widely championed climate policies could be derailed by the whims of Washington and politicians distracted by something else. It happened in 2006, when George W. Bush's promise to wean America off its "addiction to oil" quickly got shelved as the nation's attention turned ironically to a war fought in part over oil reserves in Iraq. In 2009, the most comprehensive climate bill of its time, the Waxman-Markey legislation known as the American Clean Energy and Security Act fell apart in the Senate after the Obama administration and Senate Leader Harry Reid became sidetracked and consumed with passing healthcare reform under the Affordable Care Act, better known as Obamacare. And as environmentalists and other proponents of climate action had just witnessed in horror and disbelief, even when they could pass meaningful environmental regulations, they could be thrown out on the whims a new president and his party, as Donald Trump and Congress proved by rolling back more than one hundred Obama-era climate initiatives, including the most important global agreement ever on climate change (the Paris Agreement); the most far-reaching regulations to help dirty power plants to transition to clean energy (the EPA's Clean Power Plan); and the most meaningful attempt to clean up the nation's transportation system (the EPA's and Department of Transportation's mileage and emissions standards).[24]

In 2021, COVID-19, a global economic meltdown, the worst racial strife since the 1960s, and myriad other crises were already threatening to steal all the attention and suck all the air out of the room in Congress and at the White House, once again relegating climate change as something that could be dealt with later.

Biden and team knew they couldn't let that happen.

Chapter Six

A Crisis of Crises

Joe Biden bounded onto the low stage of the Eisenhower Executive Office Building auditorium with a quick step and an air of positivity that belied his age and the fact that the country was struggling with simultaneous crises of previously unfathomable proportions. Addressing the select group of 130 or so business executives, state and federal leaders, union bosses, and journalists seated before him, he was, as usual, a bit long-winded and rambling. At different times, Biden quoted his late wife, his late mother, his late grandfather, and the late eighteenth-century essayist Samuel Johnson ("There is nothing like a hanging to focus one's attention"). Biden's homespun banter and ease with the crowd ran contrary to the state of the country.

"I think I can say without fear of contradiction, no president has ever entered office with as many crises sitting on his desk as the day he walked into office," he said. "And I've been here for eight presidents."[1]

That was in December 2009. Biden was referring to his boss, Barack Obama, at the first-of-its-kind White House Jobs Summit. It had been less than a year since the Obama administration took office, and the country was struggling with the Great Recession, war in Afghanistan, the swine flu pandemic—and if all that wasn't enough, Somali pirate attacks in the Indian Ocean.

A dozen years later, it was Biden taking office as president amid the joint crises of another recession, unprecedented racial and political unrest, the COVID-19 pandemic, and the most existential crisis of them all—climate change. By Inauguration Day 2021, COVID-19 had killed four hundred thousand Americans. The economic cratering that came with the pandemic caused even more widespread suffering, with the economy putting an estimated 9.4 million Americans out of work throughout the course of 2020, nearly double the job losses back in 2009, when Obama and Biden took office together to try and bring the country back from the Great Recession. Cities from Philadelphia to Portland, Oregon rang in 2021 not with street parties or midnight countdown choruses but were scarred with graffiti and boarded-up windows, the empty echo of *Black Lives Matter* chants from massive protests a few months earlier still ringing in the ears of some instead of the harmonies of *Auld Lang Syne.* And then there was the January 6, 2021, attack on the U.S. Capitol and the heart of democracy by hundreds of angry insurrectionists with names like *BigO*, *Baked Alaska*, and *QAnon Shaman*, who were whipped into a frenzy, directed to the Capitol, and told by the sitting president, Donald Trump, to "fight like hell" to try and stop Congress from ratifying election results showing he had resoundingly lost to Biden in the presidential election.

Looming in the background of all of this was the crisis that couldn't be cured with a vaccine or repaired like a vandalized window: The climate crisis.

When Biden took the oath of office on January 20, 2021, the East was still recovering from the busiest Atlantic hurricane season on record.[2] The West was still reeling from the record wildfires of the previous year. Forecasters predicted the year ahead—Biden's first in office—would be just as bad if not worse. But forecasters couldn't have predicted that Hurricane Ida in August 2021 would wallop cities and cause billions of dollars in damage stretching from poor St. John the Baptist Parish in rural Louisiana all the way to Seventh Avenue in midtown Manhattan, or that the Dixie Fire that started in California in July 2021 would over three months' time turn into one of the biggest conflagrations in the nation's history.

Overshadowing everything was COVID-19, which took and upended lives and shredded our economy like a hurricane in a hospital ward. Unexpectedly, the pandemic with its size, ubiquity, and unprecedented impacts also provided a rare perspective into the far-reaching costs of something as equally big: Climate change. According to the Congressional Budget Office, the virus and economic shutdown will cost the U.S. economy about $800 billion annually for the next decade, putting into perspective the expected $500 billion annual costs of climate change. The pandemic reduced the U.S. GDP by about 3.5 percent in 2020—the largest amount since World War II but only half as much as the decline that reinsurance company Swiss RE projects for the United States by 2050 if climate change is left unchecked.

People of color and low-to-moderate income communities were hit hardest by COVID-19, just as they are by climate change. Black, Hispanic, and poor Americans were three times as likely to die from COVID-19 and four times as likely to be hospitalized by the disease. Similarly, Black and Hispanic communities on average are exposed to more than twice as much air pollution as white communities. Storms and sea-level rise are most costly to poor communities and communities of color, especially in coastal states like Mississippi, Louisiana, Texas, and the Carolinas. The South and the Midwest are expected to be hardest hit economically from climate change, due to declining agricultural yields and extreme temperatures that will reduce worker productivity and also cause heat-related illnesses and death, according to a 2020 study by the World Resources Institute.[3] In the Atlantic states and Louisiana, sea-level rise alone is expected to cost as much as 1.3 percent of gross state product by the end of the century. Poorer communities and communities of color also are less likely to see the economic benefits of climate action since about 75 percent of all clean-energy workers are white. You don't see many Teslas or solar panels in places like hurricane-hit Bertie County, North Carolina, where Blacks make up 60 percent of the population, or along coastal Louisiana, which suffered through nearly twenty hurricanes and major tropical storms since the beginning of the 2020s, including three weeks in 2020 when it was under hurricane threat almost every single day while also in the middle of the COVID-19 pandemic.

Along with all its surprises, COVID-19 also delivered hope and lessons for a better, cleaner future that can also come with climate action. From Los Angeles to New York City, air pollution declined by unprecedented levels in just a matter of weeks as automobile traffic fell by 50 percent and air travel fell by even more. Factories, companies, and retail stores were forced to shut their doors, causing energy usage to plummet by as much as a fourth. By the end of 2020, U.S. carbon emissions were down 12 percent, the biggest decline of any country in the world, albeit one that proved temporary.[4] While 2020 was a year filled with dire historic precedents, it also was the first year in history that renewable energy generation outpaced coal-fired generation in the United States. Our air, our rivers, our lakes, and our oceans were suddenly cleaner. Wildlife began to reappear almost magically in some places. For the first time on a wide-scale basis, we could actually witness what cutting carbon emissions, reducing pollution and traffic, and shifting to clean energy could do to our world, like watching a flower perk back up after a good watering. COVID-19 came with death, economic destruction, and uncertainty. But the Great Pause that came with it included an unexpected pause in mankind's pressure on the natural world that had been steadily building since the Industrial Revolution. We suddenly rediscovered what life could be like in a cleaner environment, something society collectively remembered but had begun to dismiss as part of our distant past. We realized what was possible by reducing our previously insatiable demands for electricity and fossil fuels and got a glimpse of what it could look like to run our economy on cleaner, renewable energy. In short, the pandemic of a scale some thought was impossible in the modern world came with a side effect that was just as unexpected: A vision of what is possible.

It was the sort of vision that Joe Biden was looking for to set himself apart from other Democratic candidates and ultimately Donald Trump during the presidential campaign of 2020. In a normal election in a normal America, any one of the crises rocking the country on a daily basis during the summer of 2020 should've been enough for Biden to easily and definitively trounce Trump in the polls and on election day.

But 2020 wasn't a normal year, and Trump wasn't a normal candidate. He downplayed the COVID-19 pandemic by blaming it on China and suggesting it would magically go away. He pinned the racial strife engulfing the country on Antifa and Democratic politicians, and pretended the economy was fine just because the stock market was high. And when it came to the climate crisis and the cost it was inflicting on the economy, he brushed off the record wildfires, hurricanes, and storms that had been worsening his entire term as if they were just a spate of bad weather, continuing to deny climate change was happening while simultaneously rolling back limits and regulations on the main causes of it—fossil fuels, drilling, mining, and carbon emissions.

Biden saw it differently. What he saw from the campaign trail in 2020 was a confluence of opportunities to address so many of the problems he had tried to address during his lifetime in politics and public service. The racial strife that engulfed America following the killings of George Floyd, Breonna Taylor, and others that year raucously revealed to the entire country the sort of racial inequity that Biden had first seen as a white teenage lifeguard in the predominantly Black south side of Wilmington, Delaware, and later as a member of the New Castle County Council pushing to expand public housing for poor Black families. The fast and deep economic collapse spurred by COVID-19 that shuttered businesses, sent unemployment soaring, and ravaged state and city budgets across the country was eerily similar to what he saw as vice president and overseer of the 2009 American Recovery and Reinvestment Act. And the hellish fires, nonstop hurricanes, and rising costs and concerns about climate change were the sort of thing that scientists had predicted and he had warned about as a member of Congress back in the 1980s.

Biden knew all of these disparate but related crises were a clear and present danger to America and its future. But he also knew—better than most, given his unique experience—that crisis comes with opportunity, and with multiple crises come multiple opportunities. So as his political challengers fumbled trying to explain how they'd deal separately and individually with each of the myriad problems facing America in 2020, Biden cooked up a different game plan. He would tackle the crisis of crises all at once and together—with climate at the center of all of it—in the most effective way any president can quickly implement monumental changes: By leveraging the almighty power of the American economy.

Biden knew that a president's economic policies both reflect and impact the country more than anything else a president can do. Long-term presidential economic policies, at least prior to President Trump, aren't typically just about moving stock markets, or even about creating the most jobs in the shortest amount of time. They're about leveraging the economy and the government's checkbook to shape American society and address bigger, interrelated issues facing the country, in ways that can leave a lasting legacy. President Franklin Roosevelt's New Deal was designed to get the country out of depression, but it was also a blueprint to build the infrastructure needed to take America into a new era of global competitiveness and dominance. President Lyndon Johnson's Great Society had at its core the War on Poverty, but that plan and that war were shaped by battles aimed at ending segregation, advancing civil and voting rights, providing healthcare to the elderly, and revitalizing schools and inner cities. During Biden's days in the Senate, *Reaganomics* and President Ronald Reagan's regulation-cutting, trickle-down approach to the economy changed the way Americans looked at capitalism and big business, fueling the laissez-faire "Greed is Good" America of the 1980s. Later, *Bushanomics* and President George W. Bush took tax cuts and deficit spending to new levels, paving the way for consumers to do the same and making debt a critical cog in the machine that is the American economy. *Trumpanomics* was rooted in Americans' fears of foreign competition—whether from migrants from Mexico or trade wars with China—that Donald Trump eagerly exploited to push through tariffs and undo trade agreements.

Bidenomics became *climatenomics.* In one way or another, the new president's plans to confront the crisis of crises he inherited on Inauguration Day 2021 all revolved around climate action. The best way to restart the economy, Biden surmised, was by investing in the fastest-growing jobs around, namely clean-energy jobs. According to research from E2, clean-energy jobs grew at nearly twice the rate of the overall economy between 2015 and 2019. At the end of 2020, clean-energy jobs paid about 25 percent more than the median national wage (about $23.90 per hour vs. $19.14).[5] Clean-energy jobs are available to Americans regardless of education level, from entry-level construction jobs that don't require a college degree to high-level engineering R&D and finance positions. They aren't dependent on geography or geology, since clean-energy jobs exist in every state and 99 percent of U.S. counties and the sun shines and wind blows everywhere. And they're more likely to come with health and retirement benefits or be unionized than most other jobs. In a nutshell, Biden knew they were the kind of jobs that any American would be happy to have. But Biden also knew investing in clean energy would do more than just create jobs in another Great Recession. Expanding clean energy also meant reducing carbon emissions, which also meant reducing the threats of climate change and slowing the acceleration of costs that come with the hurricanes, wildfires, droughts, or floods of the future. Investing in clean-energy research and development could seed private investments and innovation in the marketplace, just like earlier government R&D investments led the way for economy-changing innovations in communications, medicine, space travel and other sectors. Launching new initiatives to upgrade the country's century-old electric grid and expand regenerative agriculture, electric vehicles, and energy storage systems could open new markets for farmers, utilities, fossil fuel companies, and other incumbent businesses, bringing them along into the clean-economy age and creating a path for growth instead of obsolescence.

And just like FDR's New Deal and Johnson's Great Society, Biden's decision to focus on climate and clean energy was also designed to drive other societal changes and benefits. With the government's hand on the tiller, Biden figured, he could steer more clean energy and more clean-energy jobs toward people of color, which in turn could help address the nation's inequity issues. As the world's biggest purchaser of energy, fuel, and equipment, the government could purchase electric vehicles, promote energy efficiency technology and clean electricity, and require contractors to hire union workers or at least pay union-equivalent wages and benefits if they want the business, lifting up working-class Americans along with CEOs. Reducing carbon emissions and other forms of pollution improves the health of Americans and the communities in which they live, possibly making the next pandemic less deadly. And getting our energy from homegrown sources like the sun and wind lessens our dependence on foreign oil and reduces what Defense Secretary Hagel coined back in 2014 as the "threat multiplier" of climate change.

Biden didn't come up with his climate-focused economic plans on his own, nor did he think for a minute he could implement them all without the help of Congress. While Donald Trump declared he "was the only one" who could solve America's problems, Joe Biden knew from long experience it takes the input of experts to figure out solutions to our country's problems, and it takes the collective will of government, businesses, and individuals to implement those solutions. That was especially the case given the breadth and depth of the crises engulfing the country in 2020, so Biden did what Biden had always done. Beginning in the summer of 2020, he reached out to and sought the advice of the people he had come to respect the most during his nearly fifty years in public service. He talked to John Kerry, his friend since the two served together in the Senate and who later served as secretary of state in the Obama administration. He consulted with Obama administration Attorney General Eric Holder, with whom he had first crossed paths in the 1980s, when

Biden was on the Senate Judiciary Committee and Holder was President Ronald Reagan's nominee to become a federal judge for the District of Columbia. He turned to old friend and union leader Lee Sanders, an African-American who grew up the son of a union bus driver in Ohio before becoming a labor economist and ultimately president of the powerful American Federation of State, County and Municipal Employees. He pulled in one of the nation's leading voices on climate action, former EPA Administrator Gina McCarthy, who at the time was running the environmental advocacy group Natural Resources Defense Council. And he spoke regularly with another friend, one whose opinions and friendship he had come to value more than any other: Former President Obama. Biden also reached outside his inner circle for advice. He started by connecting with his former rivals for the Democratic nomination, including Senator Bernie Sanders, Senator Elizabeth Warren, and former mayor Pete Buttigieg. Through them, he pulled in others. He listened to economist Stephanie Kelton, an expert on modern monetary theory who was an advisor to the Sanders campaign, and Sara Nelson, president of the Association of Flight Attendants and an early supporter of Warren's. He also brought into his circle of advisors political and climate firebrands such as Rep. Alexandria Ocasio-Cortez and Varshini Prakash, cofounder of the Sunrise Movement, the progressive climate group whose youthful members are closer in age to Biden's grandchildren than to him.[6]

Together, the small army of advisors had one thing in common: They all understood what a long-overdue focus on climate change and climate action could mean not just for the environment, but for the economy, racial equity, public health, and national security. They all realized it's all intertwined. The ideas they cooked up together as part of the Biden-Sanders Unity Task Force, as well as the ideas they presented individually to the incoming president, became the bricks in the foundation for what would become Biden's Build Back Better agenda. But more than that, it set the framework for an administration in which climate action is considered part of the means to an end for just about everything: More clean-energy jobs that come with better pay. Cleaner air and better health to help prevent everything from premature deaths to pandemics. Equity to ensure the jobs of the future and clean energy are accessible to all. Transportation and infrastructure that brings America into the twenty-first century. Improved national security and international relations. And of course, preventing the ruination of the world, the economy, and life as we know it.

Shortly after he was elected to the U.S. Senate for the first time in 1973, Biden made it a personal mission to learn the ropes from the more seasoned senators. For Biden, who at thirty years old was the youngest senator elected since Ted Kennedy in 1962, that basically meant everyone. Among those who made a particular impact on him, however, was the iconic Hubert Humphrey. Humphrey represented Minnesota for fifteen years before becoming vice president to Lyndon Johnson, only to return to the Senate in 1971 for another six-year term. Early in his career, Humphrey gave Biden some advice that stuck with him, as Biden recounted in his 2007 book *Promises to Keep.* "You have to pick an issue that becomes yours," the elder statesman told him. "That's how you attract your colleagues to follow you." As he moved into the White House in January 2021, it was clear the issue Biden had picked to become his was climate change. The ideas he took from his campaign trail advisors and the cabinet members he picked set the path and the promise for a "whole-of-government" approach to climate action, wrapped around the power of the economy.

The climate-is-economy plans Biden would unveil in his first one hundred days set the stage for what could be the most transformative period of the American economy in recent history, one that will reshape labor markets, stock markets, education systems, and the way every American gets their energy, gets to work, and looks at the future. The foundation for such a transformation existed when Biden took office. What he and his team intend to do is build the equivalent of a skyscraper upon it. If Biden and team are successful, millions of clean-energy jobs could be added each year for the next several years, a workforce multiplier that would be akin to the technology boom of the 1990s. In 2020, Tesla joined the S&P 500 after displacing Ford, GM, Exxon Mobil, and thousands of other companies in market capitalization. But if Tesla is America's next GM, the next Google or Apple or Microsoft will be a clean-energy company, if Biden and his team are successful. And just as oil and gas drove the stock market in the 1970s and 1980s and technology drove the market in the 1990s and 2000s, clean energy will drive the stock market in the years to come if Biden and market prognosticators are right.

Of course, none of this will be simple or easy. For starters, they realized, they had to undo the damage left in the wake of the Trump administration and its unprecedented rollbacks of fundamental environmental policies that set the country back decades in the fight against climate change. They had to work quickly with a divided Congress to pass policies that are resilient and lasting before the next election threatens to switch the politics of power again. They had to get business and labor on board. Most importantly, they had to change the American narrative to make companies, critics, doubters, and deniers understand the economic benefits of climate action far outweigh the economic costs.

In other words, they had to do a lot of the things that proved to be impossible to do in the past.

What's different is that climate change has never been as tangible an economic issue as it is now. Never before have the economic costs of climate change impacted everyday Americans in every part of the country like they have in recent years. More than 40 percent of Americans lived in a county that was struck by a climate-related weather disaster in 2021, according to the *Washington Post.* They included farmers suffering through droughts and flooding; business owners who had to shut down factories because extreme heat or weather damage and everyday Americans who had to skip work to clean up after hurricanes, tornadoes, fires and other calamities.

And never before have the economic benefits and opportunities of climate action been clearer in more places around the country, regardless of geography, geology, or politics. Blue-state California isn't the only place for solar businesses and jobs anymore; so are red states like North Carolina, which in recent years became one of the top states for solar;[7] and Iowa and Texas, which lead the country in wind energy generation.[8] New vehicles aren't just rolling out of left-leaning Michigan anymore, they're rolling out of factories in red states like Tennessee, where Nissan makes the electric Leaf and Volkswagen makes the electric Passat; South Carolina, where BMW also is switching to electric SUVs; and Texas, where Tesla builds electric pickup trucks. Energy efficiency jobs, which straddle sectors including manufacturing (think factory workers assembling Energy Star appliances, LED lighting systems, and insulation and other building materials), construction (think electricians, plumbers, and heating, ventilation, and air conditioning, or HVAC, technicians), and other parts of the economy, are thriving everywhere.

Members of Congress must now look at clean-energy investments and policies not as something benefitting just California or New York but their own constituents regardless of what state they represent. Similarly, members of Congress can no longer look at climate change as something causing problems in the Arctic or on some sinking faraway island but as something wreaking havoc on and hammering the economy in their own states.

Entering his first year as president, all of those stars lined up for Joe Biden to resurrect an ailing American economy, just like he helped do in 2009 as President Obama's overseer of the American Recovery and Reinvestment Act. But if Biden is successful, this time he'll do much more than that, and at a much bigger scale. He'll not only restart the economy in the wake of COVID-19, he'll completely transform it, making it more resilient and more equitable. He'll reestablish America's leadership on the existential issue of our time. And like no other president before him, he'll take real steps toward reducing the impacts of climate-related natural disasters and better prepare the country for the future climate disasters we now know will come.

Unlike his predecessor at the White House, Biden knew that he alone could not accomplish any of this, so the first thing he set about doing after assuming office in January was recruiting a cabinet of climate champions the likes of which has never been assembled before.

Chapter Seven

Climate Cabinet

When Senator Joe Biden drafted his Global Climate Protection Act back in 1986, one of his goals was to elevate the importance of global warming and climate change to the highest levels of government. If the country was ever going to do enough to address the growing national security, social, and economic threats of a warming planet, Biden thought, it had to get the attention of the president and the people who surrounded the president. And like with almost every other major issue impacting the entire world, he thought, the United States had to lead internationally as well. In his 1986 legislation, therefore, he called on the president to appoint a task force on climate change made up of experts from throughout the government and the private sector. The bill also called for the creation of an "ambassador at large" to coordinate the nation's climate efforts.

When President Biden in 2021 began thinking about how to turn climate campaign promises into reality, he turned back to his ideas from thirty-five years earlier. He also turned to one of his closest, most trusted confidants to serve as his "ambassador at large" for climate.

For Biden, there was simply no better person for the job as the nation's first Special Presidential Envoy for Climate than John Kerry. The stoic Bostonian had a lifetime of experience at almost every level of public and political service and had much in common with Biden. Before Washington, Kerry was lieutenant governor of Massachusetts under Michael Dukakis, the man credited with forcing Biden out of his first bid for the presidency, back in 1988. During the Vietnam War, Kerry was a Navy lieutenant and Swift Boat commander awarded three Purple Hearts and the Silver Star and the Bronze Star for valor. Just as importantly, like Biden, he was a fellow devout Catholic and someone who valued integrity, the fundamentals of democracy, and the institution of government. A year younger than Biden at the age of seventy-seven, Kerry also was someone who, like Biden, saw climate change through a grandfather's eyes and understood it to be a crisis their generation would shamefully leave to their grandchildren and their children after that if they didn't act, and quickly. As Secretary of State, it was Kerry (with granddaughter Isabelle on his lap) who signed the Paris Agreement that he and others in the Obama administration had helped draft back in 2015.[1] But long before then, Kerry was an outspoken but measured voice for climate action. His concern for the environment was stoked by his marriage to environmental activist and philanthropist Teresa Heinz, whom Kerry met at an Earth Day rally in 1990. During Kerry's own failed bid for the presidency in 2004, the couple traveled to nearly every corner of the country. One of the things that stuck with them about the places they went and the people they met was how climate change was already impacting communities across the nation, but also how business and innovation were creating new ways to address it, from electric vehicles to cheaper solar panels to better ways of building homes. In 2007, Kerry and Teresa Heinz Kerry literally wrote a book about how the environmental movement was shifting from something only for tree-huggers and left-leaning activists to include conservatives, evangelicals, and "bottom-line businesspeople."

"The idea that America must choose between environmental progress and economic growth is refuted by the 'green innovations' already on display by entrepreneurial pioneers and reflected in the vast new markets developing for clean-energy technologies," the couple wrote in their 2007 book *This Moment on Earth.* "Exciting new market opportunities are staring us in the face. There is enormous potential for us to create jobs, enhance security, improve the health of all Americans, and even make money, if only we seize the moment."

A year after the book was published, Kerry would co-author sweeping cap-and-trade legislation designed to reduce U.S. carbon emissions by 20 percent by 2020, significantly ramp up clean energy, and fundamentally change the American economy. The "Clean Energy Jobs and American Power Act" introduced by Kerry and California Senator Barbara Boxer was intended to be the Senate companion to the Waxman-Markey bill that the House had already passed three months earlier.[2] The Kerry-Boxer bill, however, would unceremoniously die before ever reaching the Senate floor, with Senate Majority Leader Harry Reid and the Obama administration unwilling to fight an uphill battle to pass it and instead going all-in with their political capital on another highly controversial piece of legislation, the Affordable Care Act, better known as Obamacare.

For Kerry, economics has always been a part of the solution for addressing climate change. Like Biden, he understood that the federal government must be at the center of any strategy to address climate change because of the size, scope, and consequences of the issue. But deploying more renewable energy and weaning the United States off of fossil fuels took money, more money than any government could throw at the problem. Private investors and industry had to be involved. During the Paris climate summit in 2015, Kerry and others on the Obama team helped convince Bill Gates and other billionaires to contribute $1 billion toward clean-energy innovation through a fund Gates would call Breakthrough Energy Ventures. But that wasn't enough anymore, Kerry and Biden both knew.

While a single "ambassador at large" may have been a good start to lead the nation's response to global warming back in 1986, by the time he entered the Oval Office in 2021, Biden was increasingly convinced that it was not nearly enough anymore. Climate change now required an all-of-government approach, he now knew, and it needed somebody to direct domestic strategy in addition to somebody who could direct America's international strategy on climate change.

As with Kerry, Biden turned to someone he knew and trusted, and someone who also happened to be another Bostonian.

Gina McCarthy spent her entire career in public service for the environment, beginning with the health department in Canton, Massachusetts, where she grew up. As the town's first health officer, she was engaged in everything from septic tank and restaurant inspections to one of the region's worst environmental threats, the discovery of cancer-causing PCBs at a nearby farm that led to the evacuation of a neighborhood and the ultimate designation of the farm as a Superfund site. In Canton, McCarthy learned that environmental issues were not just policy issues, they were people issues. Her work on the PCB issue, combined with her hard-charging but down-to-Earth style (not to mention the local appeal of her thick Boston accent) helped convince then-Gov. Michael Dukakis to appoint the feisty fighter for the environment to the state's Hazardous Waste Facility Site Safety Council. After Mitt Romney became governor of Massachusetts, she was promoted to the Executive Office of Environmental Affairs, where she served as the Republican governor's chief environmental adviser. Romney wasn't the only Republican governor McCarthy worked for: In 2004, Gov. Jodi Rell named her commissioner of the Connecticut Department of Energy and Environmental Protection. "That was one of the first jobs I had applied for professionally since leaving [the Canton Health Department]," McCarthy told her hometown paper, the *Canton Citizen.*[3] In Connecticut, McCarthy oversaw the state's efforts to introduce a greenhouse gas cap-and-trade program that would later help lead to the Northeast's Regional Greenhouse Gas Initiative (RGGI), which in 2009 became the nation's first mandatory market-based emissions program, now involving a dozen states. McCarthy's leadership in the Northeast captured the attention of the Obama administration, and in 2009, she became assistant administrator for the EPA's Office of Air and Radiation before Obama named her the thirteenth administrator of the EPA in 2013.

When Biden started looking for a domestic climate chief, McCarthy immediately came to mind. He didn't think he could get her. Less than a year earlier, she had become president and CEO of the Natural Resources Defense Council, where she had pledged to stay on the job for at least a few years. Besides, McCarthy already had her fill of living in Washington, and, at age sixty-six, was looking forward to some stability in her own life with husband Kenneth McCarey, her three grown daughters, and a new granddaughter, all of whom lived in greater Boston. But true to her Irish roots, McCarthy isn't one to sit on the sidelines. She'd rather pull up what she once referred to as her "ass-kicking boots" and get into the fight when it comes to saving the environment. After the Trump administration's dismantling of Obama-era environmental policies—including the signature policy she enacted as EPA administrator, the federal Clean Power Plan—she couldn't resist joining Biden, whom she knew was committed to real action on climate. She accepted the position as the nation's first White House Climate Advisor in January 2021.

"It was one of the hardest decisions I've ever had to make," she said at the time. "But when the president of the United States calls and asks you to do something, you sort of have to do it."

Besides weekly cabinet meetings with the president, it's surprisingly rare that the heads of all twenty-plus government agencies get together at the same time. Biden held his first cabinet meeting on April 1, 2021. But two months earlier, on February 11, McCarthy summoned every cabinet member (or top deputies of agencies whose secretaries had not yet been appointed or confirmed) together for the first-ever meeting of the National Climate Task Force. One of the first things she talked about at that historic February meeting was how to restore U.S. credibility on climate action in the long wake of the Trump administration.[4] She also asked cabinet members and top deputies to speak up and share exactly how they were prioritizing climate action across all of their decision-making. In doing so, she subtly put them on notice that she and the president planned to hold them accountable for ensuring every agency across the whole of government, from the Department of Defense to the Housing and Urban Development Department, was making climate a top priority.

It's tough to find anyone more passionate about clean energy than the woman Biden picked to lead the Department of Energy, perhaps the most important agency for the ongoing transformation of the nation's energy sector, and with it, the American economy. When she was governor of Michigan, Jennifer Granholm witnessed first-hand the power of the federal government to shape the private sector and save the economy after the Obama administration and Congress bailed out the auto industry with the American Recovery and Reinvestment Act during the Great Recession. She later watched wind farms spring up on the shores of Lake Michigan and Lake Huron for the first time and heard how federal weatherization projects were helping cut power bills for cash-strapped Michiganders in low-income communities in Detroit and elsewhere. Almost overnight, she became a stalwart champion of clean energy, and it changed her future. After her second term as governor, the former prosecutor who was also Michigan's first female attorney general didn't go back into law or follow the traditional route to a lobbying firm. Instead, she became an advisor to the Pew Charitable Trusts' clean-energy program and moved to California, where she was raised, and also joined University of California, Berkeley's public policy school. There, she started the American Jobs Project, an initiative to expand clean-energy jobs through public policy. She also took on a bigger public persona, becoming host of the "War Room with Jennifer Granholm" at Al Gore's Current TV network, and later became a regular commentator on politics, the economy, and climate change at MSNBC and CNN.

Granholm wasn't at the top of the list when Biden was considering picks to run the Energy Department. Especially after the antiscience Trump era, many in the environmental and business community were looking for somebody with solid science and academic bonafides to run DOE again. Under Trump, former Texas governor Rick Perry, whose academic claim to fame was that he was a Yell Leader cheerleader at Texas A&M, ran the agency despite the fact that he admitted he didn't fully understand what it did and had publicly vowed to shut it down before Trump handed him control of it. Before Perry, DOE was run by former MIT physics department chair Ernie Moniz, and before that by Nobel Prize–winning physicist Stephen Chu, both of whom understood and championed the opportunities and importance of clean energy. Biden considered going back to a science-focused DOE secretary (Moniz was a consideration, as were several other leading academics), but the president realized if he wanted to sell the climate and clean-energy ideas his team was developing, he needed to have someone who was as affable and appealing to the public as they were attuned to the latest in nuclear fission and rare Earth elements. Granholm fit that role. Shortly after her Senate confirmation in February 2021, she took to Twitter. "I'm obsessed with creating good-paying clean-energy jobs in all corners of America in service of addressing our climate crisis. I'm impatient for results. Now, let's get to work!"[5] A few months later, she joined the equally feisty McCarthy in a *Thelma and Louise*-styled video; the two grandmothers tooled around D.C. together in a government-issued Chevy Bolt. "I'm happy I'm driving, because this friend of mine gets really excited when she talks about electric vehicles and transportation," McCarthy says. On cue, Granholm excitedly explains how she leases an EV as well as the solar panels on her home. "So I drive on sunshine, girl," she tells McCarthy.[6]

It's up to Department of Transportation Secretary Pete Buttigieg to help figure out how to get more electric vehicles on American highways and get more grandmas and other Americans to use them. Like Granholm, Buttigieg was initially an unexpected pick by Biden, especially given that his only political and policy experience before running for president himself was as the mayor of South Bend, Indiana. But if anybody in the Biden cabinet will live long enough to see the worst impacts of climate change—or the ultimate results of the administration's climate action plans—it might be Buttigieg, who at 38 became the youngest secretary of transportation. Along with serving as a small-town mayor and a Navy reservist, he spent three years working as a consultant for McKinsey & Co., where he advised private and public clients, including the EPA and the Energy Department. During the 2020 presidential primary, Buttigieg burnished his climate bonafides, especially with younger voters. He spoke often about "the serious business" of climate change, and how it impacted places like South Bend, which weathered two 100-year floods within a few years of each other. As a candidate, he promised to require all new vehicles to be net-zero by 2040, ten years sooner than his boss Joe Biden, and not only cars and trucks but also buses, trains, ships, and aircraft.

"This doesn't have to be antibusiness," then-candidate Buttigieg told Bloomberg News in November 2019.[7] "What I want to do is recruit businesses to ensure they're doing the right thing and to recognize how private-sector growth is a big part of how we reach the clean-energy economy that we need."

Like Buttigieg, Biden's pick to run the Environmental Protection Agency was a young newcomer to Washington who also represented a different generation from a different part of the country with different but not dissimilar experience with climate change. Michael Regan, 45, distinguished himself as Secretary of the North Carolina Department of Environmental Quality before Biden tapped him to run EPA. In Raleigh, Regan saw first-hand how smart clean-energy policies can create jobs and drive economic growth. With nearly 100,000 clean-energy workers, Regan's home state also quickly became a leader in clean-energy jobs in the Southeast, leapfrogging its neighbors despite the fact that the sun doesn't shine any brighter in North Carolina than it does in Georgia or South Carolina or Virginia. In 2018, Regan would help draft a sweeping executive order signed by Gov. Roy Cooper to reduce the state's greenhouse gas emissions by 40 percent, cut electricity use in state-owned buildings by 40 percent, and get at least eighty thousand electric vehicles on the road.

Regan wasn't Biden's first pick to run the EPA. He wanted another female climate superstar for the role—powerful California Air Resources Board chairwoman Mary Nichols, respected not just in the United States but also internationally for her leadership on climate change.[8] But Biden's off-and-on courtship of Nichols, affectionately known as "The Queen of Green," was ultimately derailed over concerns over her history on environmental justice issues.

Biden wanted to leave no doubts with environmental justice advocates about equity and inclusion. So while he picked Regan as only the second Black administrator of the EPA (and the first Black male) and Buttigieg as the youngest and first openly gay secretary of transportation, he picked former U.S. Rep. Deb Haaland of New Mexico as the first Native American to run the Department of the Interior. Haaland, a member of the Laguna Pueblo, had a rich history fighting for environmental justice and the environment even before she came to Congress in 2019. As the business manager for the gaming company Laguna Development Corp. in New Mexico, she pushed her corporate bosses to develop and stick with a sustainability plan. Later in Congress, she worked to advance environmental justice issues for Native Americans nationwide and expand clean energy on Native American lands. Just before being tapped for Interior, Haaland introduced legislation that would require a 50 percent increase in federal investments in clean-energy research and development for everything from geothermal to battery technology. "This is good for our planet, and good for taxpayers," she would say. In picking Haaland, Biden also knew she had a great story to tell about her home state. In 2019, New Mexico became the third state (behind only California and Hawaii) to pass legislation to ultimately get 100 percent of its energy from clean, renewable sources.[9] A year later, the state passed a strategic plan to invest in, attract, and develop more clean energy. The legislation was working even before it became law: More than $80 million worth of new wind energy leases were signed in New Mexico in 2019 and 2020.

One of the most powerful and consequential players when it comes to the reshaping of the American economy to address the economics of climate change isn't at EPA or Energy or Interior. She's at the building right next door to the White House, at 1500 Pennsylvania Avenue, the headquarters for an agency that traditionally had next to nothing to do with climate change, clean energy, or the environment—the U.S. Treasury Department.

There are few people, if any, who understand the American economy and the levers the federal government can push and pull to affect change better than Treasury Secretary Janet Yellen. There are also few people who understand better than she does the connection between the impacts of climate change and the benefits of climate action to the U.S. economy. Before Biden named her the nation's seventy-eighth secretary of the U.S. Treasury, Yellen served in just about every other top role responsible for the management of the nation's finances and its economy. *Forbes* consistently names her among the World's Most Powerful People, putting her amid the ranks of Pope Francis, Vladimir Putin, Bill Gates, and Angela Merkel.[10] In the Obama administration, she was chair of the Federal Reserve from 2014 to 2018 and vice chair for the four years before that. During the Clinton administration, she chaired the White House Council of Economic Advisers. In between, she was president of the Federal Reserve Bank of San Francisco, the nation's biggest in terms of number of states and people served. She also taught economics at Harvard, Berkeley, and the London School of Economics. She served as an economic advisor to the Congressional Budget Office and the National Science Foundation. And somehow along the way she also managed to nurture a forty-year marriage to—of course—a Nobel Prize–winning fellow economist she met over lunch while the two worked at the Federal Reserve in the 1970s and raise a son, who—of course—is also an economics professor. For Yellen, it would be accurate to say her life is all about economics.

It also would be accurate to say that Yellen has been warning for decades about the economic costs of climate change and the economic benefits of climate action. For years, she has pleaded for policymakers and lawmakers to address climate change because she knows it is just as big a threat to the U.S. economy, if not bigger, than inflation, trade imbalances, unemployment, or even global pandemics. She first began to grasp this while serving as chair of President Clinton's Council of Economic Advisers, during the December 1997 United Nations conference in Kyoto, Japan, that would lead to the most sweeping climate policy of the time, the Kyoto Protocol. At the time, Republicans who controlled both houses of Congress were seeking to eviscerate Clinton and anybody in the administration who had anything to do with the treaty, which they said would ruin the U.S. economy and life in America as we knew it.

"I remember preparing my first speech on climate change not long after the conference," Yellen recounted twenty-four years later at the July 2021 Venice International Conference on Climate.[11] "This was March of 1998. The U.S. Congress was holding hearings on the agreement, and the Clinton Administration needed a White House official to testify. I raised my hand."

The hearings were before the special House Committee on the Economics of the Kyoto Protocol, a kangaroo court created by Republicans and led by Rep. Benjamin Gilman of New York, at the time one of the most outspoken critics of President Clinton in Congress. Before Gilman and other lawmakers,

Yellen acknowledged it was hard then to quantify the economic costs of climate change or the economic benefits of climate action. That was before hurricanes and wildfires were causing billions of dollars of damage each and every year. It was before solar and wind energy became commonplace, or electric cars became reality. In fact, during her testimony, she referenced the recent Detroit Auto Show, where GM had just unveiled a prototype "hybrid" vehicle that could get 80 miles per gallon. She also noted how "VCRs and TVs" consumed about $1 billion a year in electricity even when turned off. Though accurate modeling wasn't commonplace then, the economics were clear, Yellen told Congress in 1998. "The president has said that we can work to avert the grave dangers of climate change, while at the same time maintaining the strength of our economy," she said. "I agree."[12]

Two decades later, as treasury secretary under Biden, Yellen was tapped to lead efforts to do what she told Congress was hard to do back in in 1998: Figure out how to assess and address the economic costs of climate change to better insulate the U.S. economy from its "grave dangers."

Since its founding in 1789, the fundamental duty of the U.S. Treasury has been relatively simple and clear: It manages the government's finances by taking in its revenues (through its oversight of the Internal Revenue Service) and printing and distributing its currency (through its oversight of the Bureau of Engraving and Printing). Over the centuries, its duties have expanded to help oversee the health of the nation's economy by regulating its banks (through its Office of the Comptroller of the Currency) and preventing financial crimes (through its Financial Crimes Enforcement Network, or FinCEN). Today, though, one of the biggest threats to the health of the nation's economy and its financial institutions are not panicked bank runs or robberies or money laundering or counterfeiting. It's climate change.

In April 2021, at the direction of Biden, Yellen embarked on one of the most fundamental shifts in the Treasury Department's duties. She directed the agency to create more opportunities to mobilize climate-focused investments, internationally through its role with the World Bank and domestically through its Community Development Financial Institution (CDFI) and other bureaus. She directed her staff to figure out how to leverage economic and tax policies to drive more investments in climate-resilient infrastructure. And to oversee it all, she created a dedicated Treasury "Climate Hub" and appointed a "Climate Counselor"—John Morton, former Obama national security advisor for climate issues—to run it. For the first time in its 232-year history, the management of the nation's economy and finances was inculcating climate change into its duties in a very large way.[13]

Just as importantly, Yellen decided to leverage another of her even more powerful but lesser-known duties to address the economics of climate change.

In addition to managing the nation's finances, the U.S. treasury secretary since 2010 also has served as the chair of the Financial Stability Oversight Council (FSOC), the powerful but little-known agglomeration of the nation's top financial minds that was created by the 2010 Dodd-Frank Act to try and prevent another Great Recession–like meltdown of the nation's financial sector. In addition to the treasury secretary, the council is composed of ten voting members (and five nonvoting members) including the chair of the Federal Reserve, the chairs of the Security Exchange Commission and the Federal Deposit Insurance Corporation, the Comptroller of the Currency, and others. In July 2021, Yellen announced she and the council will lead a new and far-reaching effort to assess the risks of climate change to the nation's financial system.[14] The effort is the first step to creating sweeping regulatory changes across the FSOC agencies that, among other things, will require banks, investment firms, pension funds, and other financial institutions to incorporate and disclose the risk of wildfires, flooding, drought, and storms to investors' saving accounts, 401(k)s, and pensions. The goal is to protect financial institutions (and ultimately, consumers) from making risky investments in, say, oceanfront condos and office towers or housing developments in wildfire-prone areas—or more broadly, fossil fuel companies struggling or unwilling to make the transition into clean energy. The regulations will ultimately have the additional benefit of pushing financial firms to divest in fossil fuel companies and invest in climate-friendly clean-energy companies and climate-resilient infrastructure projects.

"Our pensions, our savings—our future livelihoods—depend on the financial sector to build a more sustainable and resilient economy," Yellen said in May 2021.[15]

It's not just that the vast majority of Biden's top cabinet members are longtime climate hawks and environmentalists who have been itching most of their lives for the chance to enact meaningful climate policy. Biden's bench of climate experts and clean-energy superstars runs deeper than any administration in any country in any part of the world.

Serving as McCarthy's deputy is Ali Zaidi, whom McCarthy met during her previous tour of duty in Washington and got to know better when she ran NRDC and he served briefly as a trustee of the organization. A Pakistani immigrant, Zaidi is a Harvard- and Georgetown-trained lawyer who spent eight years in the Obama administration, where he helped oversee policy and business and other private-sector partnerships at the White House Domestic Policy Council, the Department of Energy, and the Office of Management and Budget. At OMB, he oversaw nearly $100 billion in government investments related to energy, environment, and infrastructure. In doing so, he saw first-hand the role and importance of government in spurring private-sector climate investments. After Obama left office, Zaidi went to work as deputy secretary for energy and environment for New York Gov. Andrew Cuomo, where he focused on climate finance and cleaning up the state's transportation sector. Serious, studied, and in his thirties, Zaidi is one of the youngest top climate leaders.

At DOE under Granholm, there's Loan Programs Office Director Jigar Shah, who co-founded the solar energy company SunEdison and later clean-energy investment firm Generate Capital. Shah became even better known as the cohost of a geeky national podcast called the Energy Gang that's all about clean energy and climate change. There's also DOE Assistant Secretary Andrew Light, who was climate adviser to Kerry when he was secretary of state and is a renowned expert on India and its climate issues. Light also is husband to one of the nation's foremost environmental journalists, Pulitzer Prize–winner Juliet Eilperin of the *Washington Post*. At the State Department, there are climate and clean-energy experts like Melanie Nakagawa, another Kerry protege who previously worked as an international environmental attorney and later as climate advisor to a major hedge fund who now has the unique title of deputy assistant secretary for energy transformation. Under two-time Department of Agriculture Secretary Tom Vilsack is Deputy Chief of Staff Robert Bonnie, who was undersecretary for natural resources and environment under Vilsack in the Obama administration; before that, he was for fourteen years a leading policy expert at the Environmental Defense Fund.

The list goes on. In the Trump administration, there was hardly anyone who had a history of working on climate change in any sort of a leadership position. DOE Secretary Perry said he did not believe in climate change or climate science.[16] EPA Secretary Andrew Wheeler was an oil and coal lobbyist, and before him, Scott Pruitt's scandal-ridden tenure ended amid a raft of ethics investigations, after which he took a job as a fossil fuel lobbyist.[17] DOT Secretary Elaine Chao, wife of Republican Senator Mitch McConnell, ran her family's shipping business and also worked at the Heritage Foundation, the fossil-fuel funded group that's one of the biggest drivers of climate denialism in Washington and beyond.[18]

In the Biden administration, strong climate credentials are almost an unofficial requirement to land a top job. As a result, the president is constantly and consistently surrounded by people who have been pushing for action on climate change in one way or another most of their lives. As one Biden administration official put it: "Never have there been so many climate champions in an administration. And never has climate (action) been considered an economic tool like it is now."

Chapter Eight

The (Business) Plan to Build Back Better

Barack Obama was about to deliver his 2011 State of the Union address before a joint session of Congress. Environmentalists were in a tizzy. The president who for them held so much promise appeared to be going soft on climate, especially after the humiliating failure earlier of what had been the most comprehensive climate bill in history, the American Clean Energy and Security Act, also known as the Waxman-Markey bill. Environmental groups had tempered their expectations so much that, privately, their biggest goal was just to get Obama to even mention climate change during his speech. One environmental group went so far as to buy carbon offsets for the presidential motorcade, with hopes it would prompt the president to simply utter the words "climate change" in his speech.

He didn't.

What Obama did mention in his 2011 speech to Congress and the nation—twenty-five times, to be exact—was the word "jobs," pairing it with the words "clean energy" nearly a half-dozen times.[1] Sitting behind Obama in the House chamber that evening, then-Vice President Joe Biden couldn't help but take notice of the raucous applause whenever the president linked the words together. Biden had been a proponent for the environment for all of his nearly forty years in public office. But when it came to climate change, he saw it mostly as a social and national security issue, not an economic issue. That changed in 2009 when Obama tasked him with overseeing the American Recovery and Reinvestment Act. As the gatekeeper for the $830-billion economic stimulus program, Biden saw first-hand how government-led investments in solar and wind technology could drive massive new clean energy projects across America, in both red states and blue.

Biden had been a fan of weatherization projects in low-income communities for decades because they saved poor people money. But not until he was overseeing the implementation of the Recovery Act did he truly connect the dots that improving the energy efficiency of public housing also created jobs—hundreds of thousands of blue-collar jobs—installing insulation, better windows, and more efficient furnaces and hot water heaters. Biden knew the Department of Energy's national laboratories were driving innovation in solar and wind, but he didn't fully grasp the importance of other programs like DOE's loan guarantee program until seeing it jump-start game-changing companies like electric vehicle maker Tesla, even if there were some stumbles like Solyndra along the way. Biden understood how the massive purchasing power of the federal government's procurement programs could drive private markets. But he never fully realized the breadth of the market-changing possibilities like he did when the Department of Defense started buying thousands of solar panels for military housing and investing in advanced biofuels to blend into traditional fuel to power its airplanes and ships.

By the time it was President Biden's turn to give his first address to Congress, the connection between climate, jobs, and the economy was seared into his brain. A decade after he sat behind Obama on the House rostrum, Biden in his first address to Congress on April 28, 2021, mentioned "climate" six times and "jobs" forty-six times.[2] Most importantly, he linked the message together like no president before, riffing off the campaign line that helped get him elected. "For too long we've failed to use the most important word when it comes to meeting the climate crisis: Jobs. Jobs. Jobs." he said. "For me, when I think climate change, I think jobs." Especially after the previous four years of being beat down at every instance by the Trump administration, the environmental and clean-energy community went wild with approval.

Four weeks earlier, Biden had released his "American Jobs Plan," a sixty-two-page proposal for a new economy built on cleaner energy. The plan was part of a bigger program—the new president called it "Build Back Better"—to remake not just the nation's economy but everything from child-care to immigration to the federal tax code. The names were confusing, even at the White House. Biden's communications and outreach team wavered back and forth between talking about the "American Jobs Plan" and "Build Back Better" plan, often mixing the same proposals as part of either or both, until polling and public confusion finally convinced them in the summer of 2021 to call all of it the president's Build Back Better agenda. Still, to make things even more confusing to the general public, the president's proposals when they got to Congress were split into two pieces of legislation. One bill, in keeping with Congress's penchant for tortured acronyms, was originally called the "Investing in a New Vision for the Environment and Surface Transportation in America Act," or INVEST Act. The other was the bigger and bureaucratically and blandly titled "FY2022 Budget Resolution" (aka the budget reconciliation bill). With guidance from the White House and congressional communications teams, the INVEST Act was later renamed the Infrastructure Investment and Jobs Act, while the spending bill was renamed the Build Back Better Act.

Call them what you want, Biden's plans were big. All told, Democrats' infrastructure bill originally weighed in at $1.5 trillion. The early drafts of the Build Back Better Act totaled $3.5 trillion over ten years.[3] Between the two bills, the president was proposing the largest investments in the U.S. economy since World War II. Included in the early proposals were more than $800 billion in investments designed to boost electric vehicle production, rebuild the nation's century-old electric grid, ramp up clean-energy research and development, and make the government itself cleaner and greener by replacing federal vehicle fleets with electric vehicles and making federal buildings more energy efficient, saving taxpayers money in the long run. To pay for it all, the White House proposed a major overhaul to the country's tax codes, including a return to the 28 percent corporate tax rate for businesses that was in place before the Trump administration; ending international tax loopholes for big companies; increasing taxes for the very richest Americans; and, finally, eliminating many of the fossil fuel industry subsidies that for more than a century had given the oil, gas, and coal industries a competitive advantage over clean energy. Along with throwing overdue money at the problem of climate change, the president's plan also included truly transformational elements. Chief among them was the Clean Electricity Performance Program (CEPP), the White House's budgetary twist on a national clean-energy standard that environmentalists wanted. Under the CEPP, the Energy Department would set annual renewable energy goals for every utility in the country.[4] If the utilities missed those goals, they'd have to pay a fee to the government. If they exceeded them, they'd get a check from the government for every additional megawatt of energy they produced from renewables or other carbon-free sources.

The CEPP was designed to clean up existing electricity supplies. Another element of the Biden program addressed the future sources of electricity. The Clean Energy and Sustainability Accelerator (later renamed to the Greenhouse Gas Reduction Fund) was modeled after successful "green banks" around the country that provided seed capital for clean-energy companies to get projects up and going.[5] Under the accelerator program, the government would set aside $27 billion in funding, and work with private investors and commercial lenders to put up to $400 billion more into a pool that clean energy companies could borrow from at reasonable rates. With the government's skin in the game, banks and other lenders would be more comfortable putting their own money in, and more clean-energy projects could be built in more parts of the country, especially in low- and moderate-income communities and communities of color, the thinking went.

Lastly, Biden's plans also extended and expanded some of the most effective tools in the federal economic tool bag: Tax credits for consumers and businesses to install solar, wind, and other renewable energy and buy electric vehicles. Since 1992, developers and owners of wind, geothermal, and some biomass projects could get a tax credit for every megawatt of energy they produced under the federal production tax credit.[6] The bipartisan Energy Policy Act of 2005 allowed anybody who invested in solar energy—residential or commercial—to claim up to a 30 percent tax credit for their investments.[7] The 2005 act also established tax credits for cleaner fuel and cars that eventually reached as much as $7,500 per vehicle for electric and plug-in hybrid vehicles. Unlike with subsidies for oil, gas, and coal that were in place for more than a century, tax breaks for clean energy had to be debated and approved by Congress every few years, thanks to the fossil fuel industry's persistent lobbying. The tax credits for electric cars, meanwhile, were later limited to only the first two hundred thousand cars an automaker sold. Biden's plans were designed to level the playing field between clean energy and fossil fuels by extending the solar, wind, and other renewable energy tax breaks for ten years, and also include battery storage in the mix. It would level the playing field between fossil fuel–powered cars and trucks and electric vehicles by increasing the size of the tax credits for EVs and removing caps on sales.

In his first speech to Congress in April 2021, Biden didn't bring up the controversial CEPP or tax breaks or the federal green bank. Instead, as the country was clawing itself out of yet another spike in COVID-19 and the pandemic recession it caused, he focused on the winning message from the campaign trail, seeking to further solidify climate change as an economic and jobs issue that every American could understand. "The American Jobs Plan is a blue-collar blueprint to build America," Biden said. "That's what it is."

Biden's address to Congress was the starting gun for the House and Senate to begin the sausage-making of turning big ideas into legislation. Nobody expected it to happen quickly, and nobody thought it would be easy. That much was telegraphed the night of the speech by the senator who sat alone at the very back of the House chamber, rationing his limited applause and leaving out a side door as soon as Biden quit speaking. West Virginia Democrat Senator Joe Manchin happened to be the chairman of the Senate Committee on Energy and Natural Resources. He also was one of the biggest recipients of of campaign contributions from the fossil fuel industry. As a crucial and unpredictable swing vote for the slim Democratic majority, he was also suddenly the most powerful person in Congress.

When the sun rose the next day, Biden and team set about rallying support for the president's plan. The president headed to public events in Georgia and then Pennsylvania to sell the plan, while Vice President Harris took off for Wisconsin and other Midwestern states. En route, they and other senior White House officials put in phone calls to members of Congress. White House Chief of Staff Ron Klain, who had cut his sharp political teeth in Congress in the 1980s serving as chief legal counsel for the Senate Judiciary Committee (chaired by then-Senator Joe Biden), was dispatched to Capitol Hill to begin shoring up squishy Democrats. At the top of the list was Manchin and freshman Arizona Senator Kyrsten Sinema who, following in the maverick footsteps of her predecessor, Republican Senator John McCain, made it known quickly that she didn't support a spending package of the size Biden was proposing.

There would be plenty of political sparring and deal-making on Capitol Hill in the weeks ahead, Biden and his team knew. So even before the president's speech, national climate adviser McCarthy and the White House climate team were pushing cabinet members to get the executive branch going on the Build Back Better agenda. At the Department of Energy, Granholm announced she was turbocharging the agency's loan guarantee program, putting Shah in charge of it and instantly opening up $8.25 billion in loans to upgrade the nation's electric grid.[8] At the Department of Transportation, Buttigieg announced $180 million in investments for low- or no-emission buses and other transit vehicles.[9] EPA Administrator Regan, just five weeks on the job, awarded $10.5 million in contracts to replace nearly five hundred diesel school buses in forty states with electric and other zero-emission models, a welcome move for anybody who ever sat in or behind a diesel-belching school bus.[10] Yellen announced the Treasury Department's new climate hub and its new climate coordinator to run it.[11] And on the international front, Kerry and Biden just a week before the president's address to Congress convened a first-of-its-kind international climate summit, where they pledged to the world that the United States would cut its greenhouse gas emissions by 50 to 52 percent by 2030 and help jump-start an international climate finance plan to help smaller countries also reduce their overall emissions.[12]

It was the most ambitious and far-reaching few weeks the U.S. government has ever dedicated to address climate change. Never before had the whole-of-government collectively focused on climate in such a way—not during the Obama administration (when climate action took the backseat to healthcare and other issues); not during the Paris climate summit (when Republicans controlled Congress with the biggest majority since the 1930s); and not during any previous administration. And never before had climate and clean energy spending been proposed at the levels Biden and Congress were considering.

Just as importantly, outside the Beltway, businesses and the public began to get behind Biden's Build Back Better agenda. In September, nearly five hundred business leaders—tech company CEOs, real estate investors, solar and wind entrepreneurs, hoteliers, and restaurateurs—signed a letter from E2 urging Congress to put aside partisanship and work together to pass the strong climate components in Biden's plans.[13] Other organizations sent similar letters on behalf of hundreds more businesses and businesspeople.[14] More than three hundred other companies, including Microsoft, Mars, General Mills, Nike, and Salesforce, signed another letter urging federal lawmakers to support the Biden administration's Build Back Better proposals.[15] Even the U.S. Chamber of Commerce, one of the biggest opponents to climate action, surprised political, business, and environmental communities and issued a public statement in favor of Biden's plan to cut greenhouse gas emissions by more than 50 percent.[16] "U.S. businesses are leading the world in pursuit of climate change solutions," Marty Durbin, senior vice president of policy at the three-hundred-thousand-member business organization said. "And we see great opportunities to develop and export technologies that will help address a truly global challenge." The chamber's climate awakening was short-lived, however: While the big business lobby supported the infrastructure bill, it balked at the separate Build Back Better Act and its climate initiatives that the administration said were necessary to meet the nation's 50 percent greenhouse gas emissions-reduction commitment.[17] According to the business group, American companies couldn't afford the corporate tax increased designed to pay for it. In other words, the U.S. Chamber supported the idea of cutting greenhouse gases to help save the planet, but it didn't support businesses doing their part to pay for it. "The Chamber will do everything we can to prevent this tax-raising, job-killing reconciliation bill from becoming law," Chamber President and CEO Suzanne Clark declared in August 2021.

One of the biggest hurdles to the president's infrastructure bill and the separate Build Back Better Act turned out not to be the Chamber of Commerce or even Republicans, however. It turned out to be members of the president's own party. For weeks that turned into months, Democrats debated the contents of both bills to the near death of the legislation. Moderate Democrats, worried about the 2022 midterm elections, complained about the huge costs of the bills. Like the chamber, some moderates wanted to pass the infrastructure bill but were more hesitant about the Build Back Better Act spending package and its far-reaching social programs to expand government action on everything from climate to childcare. Virginia U.S. Rep. Abigail Spanberger, a Democrat whose district west of Richmond very narrowly voted for Biden—and for her—in 2020, bluntly described how she and other moderate Democrats were feeling after an unexpected November 2021 rout of Democrat gubernatorial candidates that included Democratic Party stalwart Terry McAuliffe in her own state. "Nobody elected (Biden) to be F.D.R.," Spanberger said of the president during an interview with the *New York Times*.[18] "They elected him to be normal and stop the chaos," and remove Trump from their television screens. Liberal Democrats, meanwhile, refused to give an inch on the sweeping social programs—extended support for childcare, paid family leave, price controls on prescription drugs—they had fought hard to include in the Build Back Better Act. They promised to hold the infrastructure bill, which the Senate had passed with a bipartisan 69–30 vote back in August 2021, as hostage until their demands were met. Ultra-left House members collectively known as "The Squad"—Reps. Alexandria Ocasio-Cortez, Ilhan Omar, Ayanna Pressley, Cori Bush, Jamaal Bowman, and Rashida Tlaib—pledged to vote no on the infrastructure bill unless the House had a simultaneous vote on the Build Back Better Act. As Ocasio-Cortez and others explained, they didn't trust moderate Democrats to vote for the more liberal Build Back Better Act provisions if they voted on the infrastructure bill first.[19] In doing so, congressional Democrats dealt the president delay after delay on his signature agenda, leaving him embarrassingly empty-handed as he left for the United Nations COP26 climate conference at the start of November. Biden had hoped to trumpet at the conference the passage of his Build Back Better Act as the symbol of a reemergence of American leadership on climate action that it could also use to help bring along other countries as well. Instead, Biden showed up at the Glasgow, Scotland, summit without any legislation in hand. As if that wasn't bad enough, while Biden was in Glasgow, Senator Manchin back in Washington called a press conference at which he chastised his fellow Democrats for holding up the infrastructure bill and once again denounced the Build Back Better Act as too expensive and fiscally irresponsible. "In all of my years of public service—and I've been around a long time—I've never seen anything like this," Manchin said.[20] "In my view, this is not how the United States Congress should operate or has operated in the past. The political games have to stop."

Privately, the embarrassment at the hands of his own party, coupled with McAuliffe's and other Democrats' losses at the polls, both infuriated and reenergized the president. He needed a win badly. His plummeting poll numbers reinforced as much. So even before touching down back in America after several days at the Glasgow climate conference, Biden was calling Democrats on Capitol Hill to both hear them out and give them what-for over delaying the legislation. If they didn't heed the lesson voters had just given them with the McAuliffe loss, he warned, it would doom not just his presidency but their political careers as well beginning with the 2022 midterms. He implored liberal Democrats and other fence-sitters to let House leadership pass the infrastructure bill and to believe the moderates' promises to pass the more left-leaning Build Back Better Act. He and House Speaker Nancy Pelosi, working closely with progressive House Democrats, took the unusual step of asking moderate Democrats who were giving the liberals the most heartburn, including Reps. Ed Case, Josh Gottheimer, Stephanie Murphy, Kathleen Rice, and Kurt Schrader, to issue a public statement committing they would vote for the Build Back Better Act in order to encourage the more liberal Democrats to vote first for the infrastructure bill.[21] Throughout the day and into the night, Biden was in constant contact with Rep. Pramila Jayapal (D-Wash.), leader of the Congressional Progressive Caucus. At one point, Jayapal called a closed-door session of her caucus and required all members to leave their cell phones outside to prevent leaks to the press. Then, on a landline, the president called and asked to be put on speakerphone, where he implored the gathered group of lawmakers to vote for the infrastructure bill.[22] Late on the night of Friday, November 5, 2021, after a nonstop day of tense negotiating, weary House members voted 228–206 to pass the Infrastructure Investment and Jobs Act. The six Democrats who belong to The Squad lived up to their threats and voted against it. But thirteen Republicans joined in voting in favor of the bill, giving the president a rare bipartisan victory and sending for his signature the biggest public works bill since President Dwight Eisenhower signed the interstate highway bill into law in 1956.

While the majority of the $1.2-trillion infrastructure bill is designated for roads, bridges, rail lines, and water and sewer pipes, it also includes more than $100 billion in investments in clean-energy spending, including the nation's biggest-ever investments to upgrade the nation's electricity grid, expand battery storage, electric vehicle charging, electric school buses and ferries, and weatherize homes.[23]

Ten hours after the House passed the infrastructure bill, Biden convened a 9:30 a.m. Saturday press conference. He stepped up to the podium visibly relieved and happy, chuckling as he cracked off a one-liner—"Finally, infrastructure week"—in a poke at his predecessor, who repeatedly held "infrastructure weeks" at the White House but never actually passed any substantive infrastructure programs.[24] After the tortuous few months of congressional wrangling, he could've been joking about his own party as well. Asked about his role in pushing fellow Democrats to eventually vote for the infrastructure bill, Biden said he had simply done what he had always done during his nearly four decades in Congress. "I've been doing this kind of thing my whole life," Biden said. "I spent a lot of time with a lot of people—in both political parties and within my party—saying, 'Look, if we move on what's here in this bill

. . . it is a game changer. The fact that it has too much of what you don't want and not enough (what you do want)—just, let's be reasonable.'"

The infrastructure bill had passed. But for Democrats in Washington, the work was only half done. Looming over everything was the bigger, even more transformative piece of legislation, the nearly $2-trillion Build Back Better Act. By then, the massive piece of legislation had grown to nearly 2,500 pages. While the CEPP was dropped at the demand of Senator Manchin, the bill still promised to transform America's economy not just through renewable energy and electric vehicles but also through government-supported childcare, lower prescription drug prices, and other social safety nets that Democrats had championed for years. It also contained a few conservative-sounding provisions, such as an increase in tax deductions for state and local taxes—the so-called SALT deduction—that primarily benefits wealthier residents of New York, New Jersey, and other states where property taxes on multi-million-dollar homes can be expensive.

In announcing the passage of the infrastructure bill, Speaker Pelosi also declared that the House had come to terms over when and how they would consider and vote on the Build Back Better Act. It would happen, she said, within the next two weeks. For Congress, that meant sometime before they left Washington on Friday, November 19, for the Thanksgiving break. Over the next fourteen days, Pelosi, flanked by two of her oldest and most trusted colleagues, Reps. Steny Hoyer of Maryland and Jim Clyburn of South Carolina, worked the phones nonstop. Their respective duties were clear: Pelosi would oversee the overall mission to pass Build Back Better, but also lead efforts to make sure what the House ultimately passed could make it through the Senate. That meant working closely with Majority Leader Chuck Schumer in the Senate, and also doing shuttle diplomacy with Senators Manchin and Sinema. Pelosi also took great care to make sure the bill was scrubbed of anything that could raise red flags with the Senate parliamentarian, which could jeopardize Senate Democrats' plans to pass it with only a simple majority instead of the usual sixty votes. Rep. Hoyer as House majority leader would make sure rank-and-file Democrats were supportive of the bill and address any last-minute objections or concerns they had about the bill. Rep. Clyburn, who as House majority whip was known for his vote-counting prowess as well as his uncanny ability to ferret out particular issues with rank-and-file Democrats, was the team statistician of sorts, making sure they had the votes they needed to get the bill passed.

Publicly, the Democratic triumvirate was emboldened by the bipartisan passage of the infrastructure bill two weeks earlier. There's nothing like a win to bring a team together, and the win on infrastructure seemed to both pull the different Democratic factions together and embolden the party in the

House and beyond. Privately, though, the threesome of Pelosi, Hoyer, and Clyburn had been around long enough to know that nothing was a given in politics. Even if Democrats were able to pass the Build Back Better Act, some congressional pundits prognosticated, it looked like it would once again take longer than Pelosi had hoped or promised, which would be yet another black eye for Biden and the Democrats. But on the morning of Thursday, November 18, a day before Pelosi's deadline, House Leader Hoyer's black SUV arrived at the U.S. Capitol and aides noticed there were two suitcases in the back. At Reagan International Airport, monitors showing the standby lists for several West Coast–bound flights included the names of many members of Congress and their key staffers. Confidence was growing that the bill would get passed before the end of the night.

One House member wasn't having any of that. Republican House Minority Leader Rep. Kevin McCarthy of California knew that if Pelosi was able keep all members of her party in line, there was no way Republicans could stop the bill, given Democrats' majority in the House. But that didn't keep him from trying to slow it down. Over the course of 8 hours and 32 minutes, McCarthy stood on the House floor and delivered a meandering monologue decrying the Build Back Better Act and anything else bad he could muster about Democrats. It would go down as the longest continuous speech in the House in modern history (breaking the record for a floor speech set by Pelosi in 2018). "Personally, I didn't think I could go this long," McCarthy finally said as he concluded his spiel at 5:11 a.m. the morning after he started.[25]

McCarthy's speech made no matter. A few hours after the hoarse Republican finally quit speaking, Pelosi called the vote on The Build Back Better Act. Democrats passed the bill 220–213 on November 19, without any Republican support. Only one Democrat, Rep. Jared Golden of Maine, voted against the bill, saying he could not support the SALT tax deduction, which he considered a giveaway to millionaires. In a press conference following the vote, Pelosi noted that "90-some percent of the bill" was written jointly by the House, Senate, and White House. As a result, she said, she had full confidence it would pass the Senate.[26]

"This bill is monumental. It's historic, it's transformative, it's bigger than anything we've ever done," Pelosi said, adding that she had "absolutely no doubt" it would pass the Senate. "The biggest hurdle was to get the bill there," she said. "The biggest challenge was to meet the vision of President Biden." At the White House, President Biden watched Pelosi's press conference on television, elated and relieved. It was the day before his seventy-ninth birthday.

Somebody else was closely watching the House drama unfold: Sen. Joe Manchin. If Pelosi had "absolutely no doubt" the Build Back Better Act would pass the Senate, she hadn't been paying enough attention to the doubts raised by Manchin, who would not just become the biggest hurdle, but the ultimate hurdle for getting the House-passed legislation through the Senate.

Pelosi, Biden and others could be forgiven to not realizing just how big of a hurdle Sen. Manchin would present. He had spent weeks in cordial and promising negotiations with congressional leadership and the White House. Biden was convinced he could win Manchin's support for the bill, and congressional leadership put their faith in Biden's experience, history and clout with the Senate and its senior members, which included Manchin.

What few Beltway insiders fully grasped was the simple fact that while Biden, Pelosi, Schumer and other congressional leaders saw their work through a national lens and their legislation as a way to fundamentally and positively impact the entire country writ large, Manchin's focus was shaped first and foremost by what mattered in the state where he was born and raised. Yes, he was chair of one of the most consequential congressional committees for the entire nation's future (Energy and Natural Resources) and a senior member of others (Appropriations, Armed Services, Veterans' Affairs) but first and foremost, Manchin was a born-and-bred West Virginian whose allegiance was always to Appalachia. West Virginia was still the No. 2 coal state in the country (behind Wyoming). Further, a year earlier, his state gave Donald Trump his second-biggest state victory (also behind Wyoming), with nearly 70 percent of the vote.

Manchin had problems with the most liberal provisions of the Build Back Better Act, including the expansion of child care tax credits. But he also fundamentally couldn't get his head around the fact that the legislation laid out a whole-scale shift to clean energy and away from fossil fuels, including coal. To address his discomfort, the White House and congressional leaders had included generous incentives to invest in coal communities; provisions to expand carbon capture and sequestration for coal and other fossil fuels and had agreed to leave in some extensions of fossil fuel tax breaks along with expansions of tax breaks for wind and solar. Negotiators also scaled back the overall cost dramatically to try and meet Manchin's concerns about the national budget and the impacts of inflation.

Yet with a pricetag still north of $2 trillion over ten years, including about $550 billion for clean energy and climate investments, the Build Back Better Act passed by the House was nearly 30 times the size of the entire annual economic output of West Virginia. That was still too much for Manchin to stomach. He also just couldn't get over the fact that he might be voting for legislation that could hurt his coal country neighbors and constituents, especially since they had voted overwhelmingly against Biden in the presidential election.

On December 19, 2021, exactly one month after the House passed the Build Back Better Act, Manchin went on Fox News to say he would not vote for the legislation. "If I can't go home and explain it to the people of West Virginia, I can't vote for it," he said. "And I cannot vote to continue with this piece of legislation. I just can't." Manchin's unexpected public bombshell rocked the White House, Congress and anybody who was involved in or followed the Build Back Better Act negotiations. A few weeks later, he reiterated his position even more clearly, telling reporters unequivocally the Build Back Better Act was "dead."

For Biden, congressional leadership—and the planet—Manchin's pronouncements were grim. The biggest, most important and most transformative climate and clean energy legislation ever proposed in America had withered and died at the hands of a coal-state senator from their own political party.

Yet there's something Washington insiders and policy focused wonks sometimes lose sight of, however. Regardless of what happens with policy, no matter how important and transformative it could be, the rest of the world keeps humming. And in America, where capitalism is king, where innovation rules and where the private-sector drives so much, Wall Street investors and Main Street businesses can matter as much as Washington in the fight to save our planet.

The Biden administration's original Build Back Better plan may be dead. But American businesses are already implementing plans that collectively will help build our economy back better in the wake of costly climate disasters, the pandemic and the decline of U.S. competitiveness in the global transition to clean energy. Utilities and big electricity users have already figured out that renewable energy is better (and cheaper) than fossil fuels. Automakers and consumers now realize that electric vehicles are better than gas-powered cars and trucks. Silicon Valley investors and Wall Street hedge funds now know that investing in clean energy, low-carbon agriculture and sustainability is better than putting their money in industries like oil, gas and coal whose futures are bleak.

The economics of climate change and clean energy are already incentivizing Wall Street and Main Street to build back better.

That said, we cannot rely solely on business and financial markets to meet the carbon emissions targets we know we must meet to prevent the worsen impacts from climate change. That won't work. We've tried that, and we've wasted too much time for too long. And regrettably, for every business leader who is intent on leveraging their company to do some good for the world, there's another who wants to simply make as much money as possible no matter what the consequences. Even when businesses try to do the right thing, they don't always deliver. Research group New Climate Institute took a hard look at 25 major companies that have made net-zero emissions pledges, including giants such as Apple, Amazon, IKEA, CVS Health, Nestle and Norvartis. The review, released in February 2022, found that only a few companies had implemented plans to actually meet those goals. Overall, the 25 companies were on track to reduce their emissions by 40 percent, not 100 percent. That's why federal and state regulation and policies must be a big and necessary part of addressing climate change, and why government must be right alongside businesses and individuals in the battle for the environmental and economic future of our planet. From railroads to highways to communications, America has never made a major economic transition without leadership by the federal government. And it won't do so with clean energy unless Congress acts.

There's still plenty of motivation in Washington and in statehouses nationwide to pass regulations limiting carbon emissions at big polluting companies while also incentivizing businesses and investors to expand clean energy. Even while declaring Build Back Better Act dead in early 2022, for instance, Sen. Manchin said repeatedly that he supported many of the $550 billion in climate and clean energy provisions contained in that legislation. And while they're unlikely to buck their party leadership, even some Republicans are supportive of tax incentives for wind and solar and investments in electric vehicles, energy storage and other innovations. Fact is, regardless of political party, it's hard to represent a wind state like Iowa, a solar state like North Carolina or an electric vehicle-producing state like Tennessee, Kentucky or Georgia without also supporting federal programs to keep those industries growing. It's also hard to reject policies that could reduce climate disasters and the costs that come with them when your state is increasingly getting battered by hurricanes, tornadoes or wildfire. The Biden administration and congressional leadership by no means gave up on passing truly transformative climate and clean energy legislation after Sen. Manchin declared the Build Back Better Act dead. They just realized they had to find other ways to do it.

Even so, reducing enough carbon and other greenhouse gas emissions quickly enough won't be easy. Meeting the Biden administration's goals—cutting emissions by 50 to 52 percent by 2030; converting the nation's entire electricity system to clean energy by 2035; switching American car buying to all electric or other net-zero vehicles by 2050—is daunting to say the least.

That said, remember that America has done much more in much less time. For perspective, it took a little more than twenty years for the country to bring electricity to 90 percent of American farms after passage of the 1936 Rural Electrification Act.[27] Within ten years after passage of the 1956 Federal Aid Highway Act, more than half of the interstate highway system—about 21,000 of the 41,000 total miles—had been built, and within thirty years, the system was mostly complete.[28] It took only eight years for America to accomplish what is considered one of its biggest feats, putting man on the moon.[29]

It's also important to remember that those American accomplishments by and large were primarily the result of massive government programs and came without a simultaneous rush of private-sector investments and intense public pressure and demand from consumers and shareholders, as we are seeing with clean energy and clean vehicles. When the private sector is involved—like it is now—things can speed up exponentially. Look at how fast the world shifted to personal computers after Apple and Microsoft got involved. Look at how quicky the world shifted to cell phones and mobile communications after Motorola started making them. And look at how quickly electric vehicles took off after Tesla introduced the Model 3 in 2017.

If there was any doubt about the future of cars in America or where the private sector was putting its money, you only had to tune into some of the most-expensive commercial spots on television during the most-watched sporting event in the country: The Super Bowl. Even though EVs made up only about 3.4 percent of U.S. car sales in 2021, six of the seven ads aired by car makers during the February 13, 2022 Super Bowl were for electric vehicles. According to Cars.com, that led to an 80 percent increase in EV page views across its site.

Beginning with the 2020 campaign, Biden liked to say that when he thinks of climate, he thinks of jobs. There's another saying he repeated often in 2021, and increasingly to critics of his climate and clean-energy agenda: "It's never, ever been a good bet to bet against the American people."

As America's oldest president, Biden has a perspective that others do not. From the electrification of rural America to the building of the interstate highway system to putting man on the moon, he witnessed and lived through every one of those previous American accomplishments.

Chapter Nine

Markets, Signaled

America's economy was on the verge of the biggest transition in its history. Progressives, seeing an urgent need to create jobs, make the nation more competitive in a growing international marketplace, and expand opportunities for more Americans, were proposing nothing less than to upend the very foundation of the country's economy. Doing so would effectively leave behind the industry that had almost single-handedly fueled America's economic growth from its very beginning, and just as the economy was finally showing signs of stability. Understandably, the radical plan to totally reshape the American economy stoked outrage and worry that riled up politicians and the public alike.

The year was 1790, and leading the call to build back a better economy in America was the nation's first treasury secretary, Alexander Hamilton.

Up until then, America's economy was almost entirely agrarian. An estimated 90 percent of Americans lived and worked on farms (compared with about 1 percent in 2020). If you didn't live or work on a farm, you probably worked in a business that either supplied a farm or sold their products. The engine of the American economy was powered by cotton, tobacco, produce, and livestock.

At the Treasury Department, Hamilton saw a different future, one with a more diversified economy, with more job opportunities for more people (including women and children) and one that would attract badly needed foreign investments and workers to the young nation. So in 1791, at the request of Congress and with the encouragement of President George Washington, Hamilton penned "Report on the Subject of Manufactures," a seminal economic proposal that at the time was as incendiary in the political and public discourse as a tweet might be from today's treasury secretary suggesting an end to the stock market.[1] Hamilton's eighty-page proposal basically suggested that if the United States was going to be competitive and continue to grow, it needed to do more than just farm. He urged Congress to use the "incitement and patronage of government" to help grow and nurture the nascent industry of manufacturing, for the betterment of the country and its economy. Back then, mass-production of goods with mechanized factories was as foreign and inconceivable as electric cars or glass panels that could make electricity from sun were a few decades ago. He proposed a series of "pecuniary bounties" and "premiums," subsidies that included tax incentives and government investments to encourage the growth of manufacturing for products such as coal, cotton, wool, sail cloth, and glass. Hamilton's "Manufactures" also suggested the need for new policies to encourage immigration of skilled workers and provide training for others to expand job opportunities. To help pay for his sweeping economic plan, he proposed tariffs on imported finished goods.

Agricultural interests, led by powerful patrons Thomas Jefferson and James Madison, pushed back hard. Supporters of American agriculture lobbied the Congress to vote against Hamilton's proposals, citing everything from the sanctity of farming to the constitutionality of federal subsidies. Jefferson, a resolute champion of the virtues of American agriculture and a resolved opponent to government manipulation of the economy, was particularly incensed, at one point writing that Hamilton's proposals to support the nascent manufacturing industry "flowed from principles adverse to liberty . . . calculated to undermine and demolish the republic."[2]

Congress never passed Hamilton's full package, although it did ultimately implement most of the tariffs he proposed (including the first of many U.S. subsidies to the fossil fuel industry, a tariff on British coal designed to help domestic coal companies). Even so, the push by the nation's treasury secretary to expand and support manufacturing, and his suggestion that the government could provide subsidies and impose taxes and tariffs to help the industry grow, wasn't lost on investors and businesspeople of the time. The market was signaled.[3] The same year as Hamilton's report, businessman Samuel Slater mechanized the first factory in America by adding a water-power wheel to his cotton-spinning mill in Pawtucket, Rhode Island.[4] Two years later, Eli Whitney applied for a patent for the cotton gin, jump-starting the American textile industry and sparking America's industrial revolution that would transform the nation's economy and turn it into an international superpower. By the time of their presidencies, both Jefferson and Madison adopted policies and positions to support the U.S. manufacturing industry nearly identical to those from Hamilton they so vociferously opposed, including the 1816 Tariff Act signed by Madison to explicitly support American manufacturers.[5]

The ability of the federal government to send market signals to shape the economy and incentivize specific industries has come a long way since "Report on the Subject of Manufactures." The government's R&D investments and other support of computing and communications innovation through its Advanced Research Projects Agency Network (ARPANET) led to the modern-day Internet and the biggest transformation of the nation's economy in recent history. Homeownership and the real estate industry in America wouldn't be what it is without government-sponsored Fannie Mae, Freddie Mac, and Veterans Administration loans and tax credits for mortgage interest. And yes, federal incentives and tax breaks for the oil, gas, and coal industry led fossil fuels to become the nation's dominant source of energy, its most profitable industry, and the engine of the U.S. economy for more than a century.

Of course, market signals only work when businesses in those markets are mature enough and ready to receive them. Jimmy Carter learned that the hard way. In 1977, with the country still reeling from the Arab Oil Embargo energy crisis, then-President Carter signed legislation creating the Department of Energy, charging the new agency with the primary duties of researching nuclear weapons and developing new sources of renewable energy and fuels so the country would be less dependent on foreign oil.[6] Carter also created the Solar Energy Research Institute, the predecessor to today's National Renewable Energy Lab (NREL). A year later, he introduced the National Energy Act, which created a package of market-based initiatives including tax breaks and other incentives for energy efficiency, renewable energy, and alternative fuels. If that weren't enough to send a strong signal to private industry to invest in and develop more solar, Carter in 1978 went so far as to declare May 3 as "Sun Day" to promote solar energy, and then the following year took it to yet another level, namely the roof of the White House, where he installed thirty-two solar panels to produce hot water for the People's House.[7] Never had there been a clearer market signal to business and industry that the federal government was serious about solar.

It hardly mattered. Despite Carter's efforts, in the thirty years between 1978 and 2008, only about 0.34 gigawatts of solar energy was installed in the United States, according to the DOE.[8] That's less solar energy than Walmart generated in 2019 from panels on top of its stores. Market signals from the federal government in Carter's time didn't work because the cost-benefit ratio of clean energy didn't work. Solar panels were still too expensive (about $40 per watt in the 1970s, versus less than $1 per watt in the 2020s), and the economic costs of climate change tied to the burning of fossil fuels weren't as clear as they are today. Forty years later, Carter would be redeemed as solar displaced coal and gas as the cheapest electricity available in many parts of the country. In 2017, the former president once again decided to install some solar panels at his home: 3,800 of them, in fact, on ten acres of his family's Plains, Georgia, farm that he leased to an Atlanta solar company called SolAmerica.[9] Since then, Carter's panels have provided about 50 percent of the electricity for his entire town.

Even if the technology is ripe, market signals don't matter if businesses ignore or kill them. In 1990, the California Air Resources Board (CARB) passed the most ambitious clean vehicle regulations of the time.[10] CARB required any automaker wanting to sell vehicles in California, the country's largest car market, to produce and sell zero-emission vehicles, such as those that run on electricity or hydrogen. Under the CARB rules, at least 2 percent of automakers' fleets had to be zero-emission by 1998, and at least 10 percent of their cars had to be zero-emission by 2003. CARB and automakers knew these goals were reachable: Earlier in 1990, automaker General Motors had unveiled the first-ever mass-produced electric vehicle in the country, the GM EV1.[11] The automaker went on to produce more than a thousand of the sleek two-seater sedans between 1997 and 1999, almost all of which were leased to residents of California and Arizona. Just like Tesla drivers decades later, passionate and loyal EV1 drivers in the late 1990s loved the vehicles for their cost-savings. They lauded their small impact on the environment. And they basked in their coolness. It looked like electric vehicles were finally on the road to going mainstream in America.

But none of that mattered. Backed and pressured by fossil fuel interests and others, GM and other automakers sued CARB to water down and roll back the ZEV regulations.[12] In 2003, GM canceled production of the EV1, took back almost every car it had leased, and crushed them all, as recounted in the documentary *Who Killed the Electric Car?* Despite the government market signal, despite the rave product reviews and the growing public demand, EVs wouldn't go mainstream in America until Tesla came along and revolutionized the auto industry—thanks in part to a modern-day "pecuniary bounty," namely a $465-million Department of Energy loan the start-up company received in 2009 as part of the American Recovery and Reinvestment Act. Nearly twenty-five years after the first doomed EV1 rolled off the factory floor, GM announced in 2021 it would follow Tesla's lead and by 2035 produce only electric vehicles.

What makes this point in time different when it comes to clean vehicles, clean energy, and climate action is that all the pieces have come together. Pundits and prognosticators—and some businesspeople—have predicted this transition for a very long time. But never have the fundamental elements of economics, climate change, and government policy—*climatenomics*—converged like they have in the 2020s.

"I've never seen where businesses, customers, governments, partners are all aligned and seeking action" like they are on climate change, Johnson Controls CEO George Oliver said at a September 2021 Business Council for Sustainable Energy event.[13] Oliver has seen a lot: Before becoming CEO of Johnson Controls in 2017, he was CEO of conglomerate Tyco and before that spent twenty years in executive roles at General Electric.

To see what Oliver and others see, start by looking at technology and costs. Unlike in Carter's days, the technology for solar, wind, electric vehicles, and battery storage is now ripe and proven. As a result, costs for cleaner energy and transportation have plummeted. Between 2009 and 2019, solar energy prices fell nearly 90 percent, and wind energy prices fell 70 percent, according to financial advisory and research firm Lazard.[14] While overall new car prices jumped by 12 percent in 2021 amid global supply shortages, the average cost of electric vehicles grew by half as much. With an average price tag of about $56,500 at the end of 2021, new EVs were cheaper than full-size pickups or SUVs and only about $10,000 more than the industry average for all cars, even before any government incentives, according to Kelley Blue Book.[15] Thanks to declining prices, a record half-million electric vehicles were sold in the United States in 2021, with EV sales jumping 72 percent in the fourth quarter of 2021 alone.[16] Prices for lithium-ion batteries, the core of electric vehicles and utility-scale energy storage, fell 85 percent between 2009 and 2019, and the average cost of a four-hour battery is expected to be nearly half as much in 2030 as it was in 2020, projections from the National Renewable Energy Laboratory show. The trendline is already clear: A 75KW battery that might go into a Tesla cost less than $10,000 in 2021. A decade earlier that same size battery cost about $33,750.[17]

Next, look at business buy-in. Unlike with GM's flirtation with electric vehicles in the 1990s, businesses are now all-in on clean vehicles and clean electricity. Beginning in 2021, every major automaker announced they were shifting to electric vehicles. They're doing so because their customers—they're not just a handful of eco-minded consumers in California and a few other states anymore—are buying in to EVs, too. UPS announced it would buy ten thousand electric vehicles and took a minority stake in a European EV company.[18] Amazon ordered one hundred thousand electric vehicle delivery trucks, and Hertz announced it would buy one hundred thousand Teslas for its rental car business.[19] Six months after Ford announced it would make an electric version of the best-selling vehicle in America, its F-150 pickup, it had 200,000 advance orders for the new F-150 Lighting truck. The response was so unexpectedly overwhelming that the company said it would have to quit taking pre-orders because it couldn't build enough electric trucks fast enough. It also announced it would double production of the Lightning and build a

US LARGE SCALE WIND DEPLOYMENT

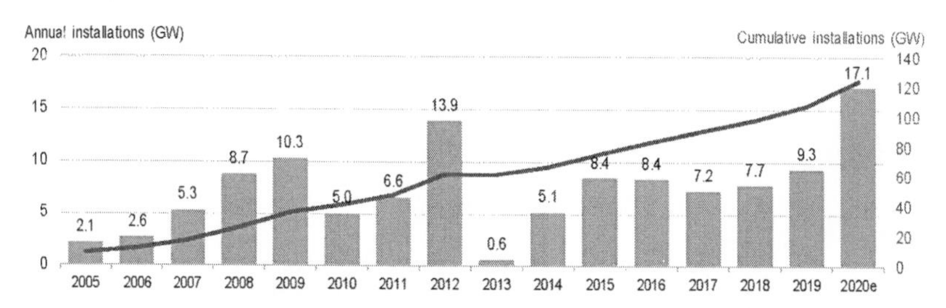

Figure 9.1. As prices decline, wind deployment continues to rise. *Source*: Bloomberg NEF, the Business Council for Sustainable Energy (BCSE), *Sustainable Energy in America 2021 Factbook.*

new truck factory and three new battery plants to just keep up with demand. Ford and other EV makers are scrambling to meet what Wedbush Securities estimates will be a $5-trillion market for EVs by 2030.[20] Automakers are also looking beyond electric batteries to clean up their products. Volvo AB, for instance, announced in 2021 that it will begin making vehicles from "green steel" made without fossil fuels, through a joint venture it has with Swedish steelmaker Hybrit.[21]

While car companies shift to electrics, electric companies are shifting to solar, wind, and batteries. NextEra, the nation's biggest electricity company by market capitalization, now operates the biggest solar fleet in the world outside of China.[22] Investor Warren Buffett's Berkshire Hathaway Energy, the holding company for MidAmerican Energy, sourced half its energy from renewable sources in 2019 and expects to be able to meet all its demand from its massive wind farms alone.[23] Even utilities in the Southeast, which have traditionally fought the hardest to preserve a coal- and gas-based energy sector, are now rapidly shifting to wind and solar and away from coal. Atlanta-based Southern Co., for instance, once operated one of the biggest fleets of coal-fired power plants in the country, with sixty-six generating units throughout the Southeast. By November 2021, it had only eighteen coal-fired plants, and company CEO Tom Fanning told Wall Street analysts it would reduce those by half by the end of the decade as part of its plans to achieve net-zero greenhouse gas emissions by 2050.[24] In Virginia, Dominion Energy in November 2021 submitted plans to build the largest offshore wind project in the United States, a 2.6-gigawatt array off of Hampton Roads that will generate enough energy to supply six hundred thousand homes, create an estimated eleven thousand jobs, and generate $210 million in economic growth.[25] "Our customers expect reliable, affordable, and clean energy, and we intend to deliver," Dominion chairman and CEO Bob Blue said in a statement about

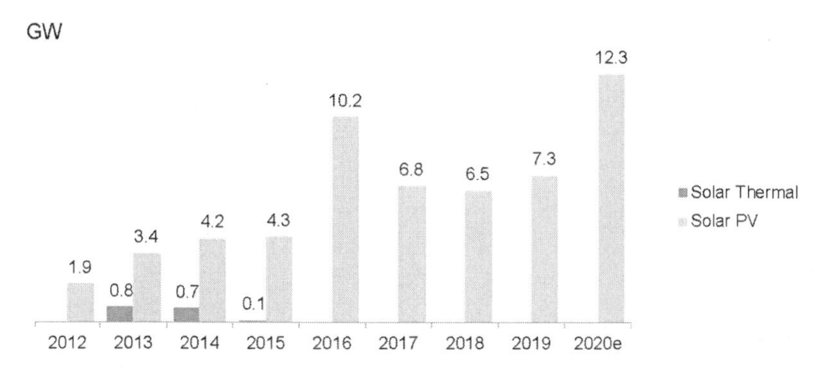

Figure 9.2. Solar deployment hit a record high in 2020 and was on pace to hit another record in 2021. *Source*: Bloomberg NEF, the Business Council for Sustainable Energy (BCSE), *Sustainable Energy in America 2021 Factbook.*

its offshore wind plans. "In addition to solar, storage, and nuclear, offshore wind is a key component of our strategy and a game changer for the Hampton Roads economy." Driving Dominion, Southern, and other companies are three things. First and foremost is the fact that clean energy is cheaper—and therefore more profitable—to produce. Then then are the government incentives to build clean energy and government regulations to stop using fossil fuels. And finally, as Dominion CEO Blue noted, there's customer demand.

In 2021, some of the largest business users of electricity—companies such as Amazon, Meta, Verizon, and Google—purchased 11 gigawatts of renewable energy, according to the Clean Energy Buyers Alliance (CEBA).[26] That was up from 9.3 gigawatts two years earlier and more than three times the amount they purchased in 2015. Nearly four hundred big companies and other major users of electricity—including Microsoft, Procter & Gamble, Bank of America, Wells Fargo, and General Mills—sourced 100 percent of their energy from renewables, according to the US EPA's Green Power Partnership program.[27]

Of course, like utilities, other businesses aren't moving to clean energy and electric vehicles to save the planet; they're doing it to cut costs. The per-mile cost of operating electric vehicles is about half the cost of gas-powered vehicles. Consulting firm McKinsey estimates businesses that deploy electric vehicles throughout their fleets could save as much as 25 percent in operating costs compared to using internal combustion engines.[28] With those types of savings, 15 percent of all business vehicle fleets will be electric by 2030, McKinsey predicts.[29] Similarly, now that solar and wind provide the cheapest power available, businesses can further cut costs by shifting to clean energy.

Data center and records management company Iron Mountain learned this first-hand. When the Boston-based company was looking for ways to reduce power costs and increase reliability for its data centers in Pennsylvania, it turned to wind energy. In 2016, Iron Mountain agreed to a fifteen-year agreement to purchase energy from the Ringer Hill Wind Farm, a 14-turbine, 40-megawatt operation south of Pittsburgh. Iron Mountain estimates it will save $2 million per year—or $30 million over the lifetime of its power purchase agreement. BlackRock CEO Larry Fink might have explained it best in his 2022 Letter to CEOs. "We focus on sustainability not because we're environmentalists, but because we are capitalists and fiduciaries to our clients," he wrote.

And then there's that market signal from the government. The Biden administration's climate and clean energy agenda is a three-pronged effort that includes regulation, legislation, and procurement. Together, it's designed to send a market signal that couldn't be any clearer.

On the regulatory front, step one for the Biden administration was to halt or roll back scores of anti-environmental policies implemented by the Trump administration. Then, it began working on new regulations. The Department of Energy, for instance, is planning to upgrade energy standards for more than 60 appliances and other pieces of equipment—furnaces, clothes dryers, freezers, light bulbs—between 2022 and 2024. While bland, efficiency standards are important: They're what fostered innovations like LED lightbulbs, high-efficiency dishwashers and refrigerators and heat pump hot water heaters. The new DOE standards will similarly encourage innovation and impact businesses ranging from appliance factories to big box home retailers all across America. New auto emissions standards announced by the Department of Transportation and the Environmental Protection Agency in December 2021 will result in car makers producing vehicles that average 40 miles per gallon across their fleets by 2026, in keeping with the Biden administration's goals of reaching 100 percent net-zero vehicles by 2050. In the past, vehicle mileage and emissions standards have pushed automakers to develop and expand hybrids and electrics, but also innovations such as start-stop technology that reduces idling and design features such as lighter materials. Similarly, the latest standards are expected to drive new innovation and growth in the auto industry. Department of Treasury requirements will for the first time push banks and investment firms to account for climate risk in the investments they make on behalf of their shareholders. The Environmental Protection Agency will also help drive the transition away from fossil fuels by enforcing standards on oil and gas industry methane missions, limiting greenhouse gas emissions by utilities and other big polluters, and tightening up how coal companies operate. The Department of Interior, meanwhile, opened vast new areas of the Atlantic and Pacific up to offshore wind energy development in 2021 and 2022, while simultaneously making it harder for oil and gas companies to drill offshore or on federal lands. The DOI's actions will bring offshore wind and billions of dollars in investments that come with it for the first time to California and other Pacific states, while greatly increasing the wind energy produced on the East Coast with new projects stretching from the New York Bight to coastal North Carolina. Off New York alone, the opening of 480,000 acres of ocean in the Biden administration's first offshore wind auction is expected to result in enough energy to power two million homes. It also will create tens of thousands of new jobs and billions of dollars investments. In January 2022, the state of New York further primed the pump, investing $500 million in upgrades to ports and supply chain infrastructure to open the way for the region's offshore wind industry.

On the procurement front, the Biden administration is leveraging the federal government's power as the world's biggest buyer of goods to create new markets and drive growth in clean energy, clean vehicles, and energy efficiency. On December 8, 2021, President Biden signed an executive order directing 100 percent of the electricity used in federal government buildings to come from carbon-free sources by 2030, 100 percent of cars and trucks used in federal fleets to be electric or other zero-emission vehicles, and for federal buildings to be net-zero emissions by 2045. The impacts of that sort of purchasing power are simply huge. The General Services Administration, for instance, owns or leases about 9,600 buildings across the country that will soon need to get all of their energy from renewable sources. Many of them also will need energy efficiency upgrades—think everything from more insulation and better windows and doors to new lighting and HVAC systems—to meet the net-zero building requirements. Meanwhile, more than 650,000 trucks, vans, and other vehicles operated by government agency fleets will soon be replaced with EVs or other zero-emission vehicles. For one example of how this will play out, look to the U.S. Army. In February 2022, Army Secretary Christine Wormuth released a sweeping climate strategy that calls for the Army to power all of its 130 installations worldwide with carbon-free clean energy by 2030 and for all of its non-tactical vehicles to be all-electric by 2035. Those shifts are huge: The Army spends about $740 million on electricity every year, and the bulk of the Department of Defense's 170,000 non-tactical vehicles can be found at Army bases around the world, currently burning millions of gallons of gas and diesel annually. Under orders from Secretary Wormuth (and ultimately, the Commander-in-Chief) the Army needs to buy tens of thousands of new electric cars, trucks and buses and also get all its electricity from solar, wind or other carbon-free sources within 13 years. Such procurement will create huge new markets for American clean energy and clean transportation companies, and along the way drive innovation, increase manufacturing capacity and ultimately reduce prices throughout the economy. In addition to creating instant sales for renewable companies, car dealers, and energy efficiency contractors, the federal government's big-ticket orders also promise to drive down prices and spur competition and innovation among private-sector contractors and suppliers that will reverberate throughout the economy.

And then there's the legislation.

The Infrastructure Investment and Jobs Act passed by Congress and signed by Biden in November 2021 includes more than $100 billion to be invested in clean-energy-related projects,[30] including:

- $65 billion for upgrading the nation's power grid so it can more handle more renewable energy and battery storage, and to make it safer and more resilient.
- $7.5 billion to roll out five hundred thousand electric vehicle charging stations—quintupling the number of EV charging stations in America in 2020.
- $5 billion for electric and other zero-emission school buses.
- $3.5 billion for weatherization projects, with a focus on homes in low- and moderate-income communities and communities of color.
- $1.5 billion in energy efficiency and conservation block grant programs.
- $750 million in advanced energy technology projects in coal communities.
- $21 billion for clean-energy demonstration projects and research hubs.

In itself, the infrastructure package included some of the biggest investments in climate and clean energy that the country has ever made. But while they are significant and will help drive private-sector growth in a big way, the infrastructure package projects are not nearly enough to meet the carbon emissions reductions needed to prevent the continued worsening of climate change and the economic costs that come with it, or the climate pledges the Biden administration made to the world. For that—and also to reap the jobs, investments and economic growth that come with a clean energy future—the country needs to invest at the levels outlined in the original Build Back Better Act, environmentalists, scientists and economists agree.

We already know how important and transformative this sort of legislation can be. The investment tax credit for solar was enacted in 2006, providing a 26 percent tax credit for installing residential or commercial solar electric systems. Since then, the solar industry has grown more than 10,000 percent—a 52 percent average annual growth rate—and generated billions of dollars in investments, according to the Solar Energy Industries Association.

Along the way, solar companies have created more than 315,000 jobs that in 2020 came with a median hourly wage of about $24.50, according to E2 research.

If Congress expands clean energy tax credits for 10 years—as outlined in Biden's Build Back Better Act—the economic benefits would be roughly 3-4 times greater than the costs, a February 2022 study by Rhodium Group shows. According to Rhodium, the clean energy tax incentives would cost the country $309 billion or less in lost revenues, but they would generate as much as $1.8 trillion in economic benefits.

Similarly, the electric vehicle tax credit, established by President George W. Bush with the 2008 Energy Improvement and Extension Act, has helped make some of the most expensive vehicles in the country among the most popular. The 2008 law established that buyers of electric vehicles could get up to a $7,500 tax credit for new vehicles, until the point at which a manufacturer sells more than 200,000 vehicles. A study by the University of California, Davis in 2017 showed that nearly 30 percent of consumers who bought electric vehicles cited the EV credit as a factor in their decision. By 2019, as the EV credit became better known, it was an even bigger influence in car buyers' decision-making. In a poll by ClimateNexus, almost 75 percent of respondents said a guaranteed federal tax rebate of $7,500 would make them more likely to consider an EV.

Government procurement, tax incentives, and other investments—"pecuniary bounties," as Hamilton would call them—are sending clear market signals pulsing through the private sector, encouraging a shift to a cleaner economy. That will only continue as the Biden administration rolls out its regulatory and administrative actions in the months and years ahead, and as Congress finds ways to refocus on climate-centric legislation. Just as government investments drove the rise of manufacturing back in Hamilton's day, and more recently drove the rise of the Internet, the private space industry and many other sectors of the economy, government investments in clean energy, clean transportation and climate resilience are once again setting the pathway and spurring on the private sector to once again drive radical economic change.

The federal government's investments, regulatory action, procurement other policies "sends a gigantic signal to the private sector," National Climate Advisor Gina McCarthy told me as part of an E2 event in late 2021.[31] "And (businesses) are so creative about making their own path forward, that generally when the government sends a signal like this, it just takes off."

"It's like releasing a boulder and sending it downhill," she said. "Over time, it just picks up speed."

Chapter Ten

From Sacrifice to Shared Prosperity

On the evening of February 2, 1977, President Jimmy Carter sat alone next to a fireplace in the dimly lit White House, wearing a tan cardigan to address the nation about the ongoing energy crisis.[1] With gas and oil prices soaring, Carter told Americans already suffering through a harsh winter they should prepare for more sacrifice. The world was still recovering from the Arab Oil Embargo. There was a natural gas shortage. During the three months before Carter spoke, nine oil tankers had either sunk, run aground, or leaked, beginning with the *Argo Merchant,* which broke apart on December 15, 1976, spilling nearly eight million gallons of oil off Nantucket, Massachusetts, and leading up to the fuel barge *Ethel H.* running aground in New York Harbor on the very same day as Carter's speech, spewing 480,000 gallons of crude into the icy harbor.[2] It was clear, Carter declared, that the United States needed a national energy policy, something every other major country already had at the time, and a Department of Energy to oversee it. Everyday Americans had to do their share, the president sternly warned. He urged his fellow countrymen and women to turn their thermostats down to 65 degrees during daytime and 55 at night. Businesses would have to make sacrifices too, he said, as would the government. Carter's sweater speech took place just two weeks after he took office, but it quickly went down as one of the worst speeches by any president ever. It was Carter's first stumble in what would be a series of face-plants trying to communicate his approach to what at the time was America's biggest crisis—the energy crisis. When Carter was ousted from office after a single term, pundits would look back at that fireside speech as the beginning of the end of his presidency.

Joe Biden's political adversaries want people to believe his clean-energy policies are Jimmy Carter's all over again: Expensive, unpopular, and not ready for prime time. Early on in Biden's presidency, big lobby groups such as the American Petroleum Institute, the National Association of Manufacturers, and the U.S. Chamber of Commerce all surprisingly declared they supported Biden's strong and sweeping climate and clean-energy policies. But within a few months, they were back to their old selves, all announcing they would fight tooth-and-nail against the spending packages and corporate tax hikes to pay for them. The *Wall Street Journal*, Fox News, and other conservative outlets opined that Biden's clean-energy plans would result in massive blackouts and put millions of oil and gas workers out of work—even though only 891,000 Americans worked in oil and gas extraction in 2020.[34] Senate Minority Leader Mitch McConnell took to the Senate floor with talking points straight from the Carter-era Republican playbook, saying Biden was "willfully throwing our people out of work, reducing our domestic energy security, raising costs and prices for working families."[5] Ohio Republican firebrand Rep. Jim Jordan didn't waste time with innuendo, tweeting in May 2021: "Joe Biden is the new Jimmy Carter."[6]

But suggesting that Biden's vision and the state of clean energy today is the same as his predecessor's from the 1970s is as dated as Carter's cardigan. In the 1970s, solar energy cost exponentially more than any other source of power. Most Americans considered wind energy as outdated as an eleventh-century Dutch windmill. Electric cars back then were as futuristic as video-conferencing or mobile phones. At the beginning of the 2020s, electricity from new solar and wind farms cost about a third of the cost of electricity generated by new gas and coal plants.[7] Energy efficiency today isn't limited to turning down the thermostat and pulling on a cardigan. It means turning on LED lights that are brighter, better, and operate more cheaply than any incandescent from Carter's days. It means using your wireless thermostat to control your high-efficiency HVAC system, your cell phone to put your heat pump hot water heater in energy efficiency mode, and your touch-screen induction stove to cook your dinner quicker, cleaner, and more safely than any traditional electric or gas-burning oven. Electric cars today are nearly equivalent in price to many gas-powered cars, but they're also 50 percent cheaper to operate, last three times longer, and you can fill up at home while you're sleeping.[8]

Clean energy is not just cheaper; it's *cooler.* In the 1960s, Lee Dorsey sang about *Working in the Coal Mine.* In 2021, one of the hottest summer hits was singer Lorde's *Solar Power.* Teslas and Wi-Fi-controlled HVAC systems have become the energy equivalent of the newest Apple iPhone or Alexa wireless home assistant. Young people graduating from school don't want to work in a coal mine; they want to work at a battery storage start-up or in clean-energy finance. Unlike in Carter's day when solar was a novelty, Americans today see solar panels sprouting from rooftops in every state. They're experiencing

LAZARD LAZARD'S LEVELIZED COST OF ENERGY ANALYSIS—VERSION 15.0

Levelized Cost of Energy Comparison—Historical Renewable Energy LCOE Declines

In light of material declines in the pricing of system components and improvements in efficiency, among other factors, wind and utility-scale solar PV have exhibited dramatic LCOE declines; however, as these industries have matured, the rates of decline have diminished

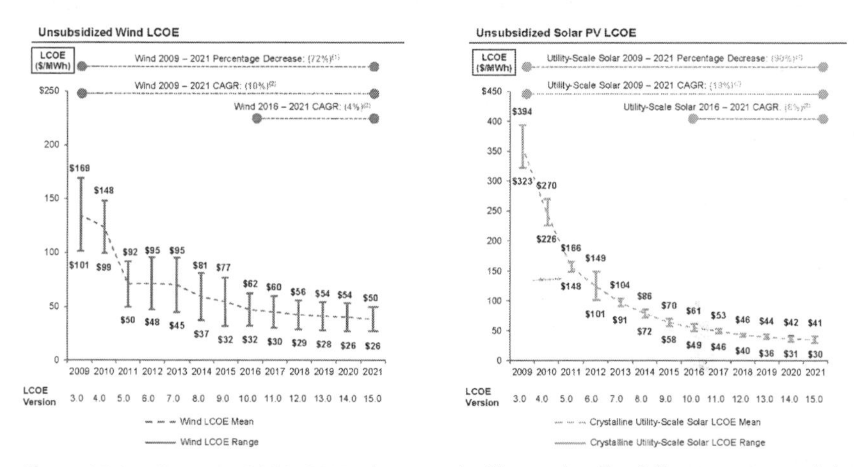

Figure 10.1. Between 2009–2017, the cost of utility-scale solar fell 90 percent, and the cost of utility-scale wind dropped 72 percent. *Source*: Lazard, "Lazard Levelized Cost of Energy Analysis, v. 15 (October 2021)."

first-hand savings from better light bulbs, HVAC systems, and hot water heaters and appliances. If they don't have an electric vehicle, they probably know a friend or family member who raves about owning one.

And they want more.

More than 70 percent of Americans polled by *Consumer Reports* in December 2020 said they were interested in buying an electric vehicle.[9] About 84 percent of respondents surveyed by Pew Research in June 2021 said they wanted more solar energy, and nearly 80 percent said they wanted more wind energy in America.[10] Two-thirds of Americans polled by Pew said the government should do more to address climate change, with about 80 percent saying it should be a national priority to develop more clean energy. And yes, Republicans and Democrats may disagree on almost everything and partisanship is worse than ever, but there's at least one thing they agree on: The need for more federal action to accelerate clean energy. A June 2020 poll by the conservative Citizens for Responsible Energy Solutions showed that about 75 percent of Republicans and Republican-leaning voters support federal action to accelerate clean energy. In the same poll, nearly 70 percent of Republicans said investing in clean energy was important to rebuilding the economy after the COVID-19 shutdown, and key to helping the United States become a world leader in "green economic development."[11] Whether they like Joe Biden or not, Republicans, like most Americans, like the idea of clean energy, and they understand transitioning away from fossil fuels can create jobs, drive economic growth, and keep America competitive.

What's also different now is the public's understanding of climate change and their own experience with it. Unlike in Carter's days, climate change is no longer something off in the future, threatening polar ice caps or some faraway Pacific island. It's happening now, and it's happening in our backyards. According to a *Washington Post* analysis, nearly one in four Americans lived in a county hit by a weather disaster in 2021.[12] Another way to look at that: About 40 percent percent of Americans lived in a county that was declared a weather-related disaster area by the Federal Emergency Management Agency. Just three years earlier, in 2018, only 5 percent of Americans lived in a county that had been declared a weather-related disaster area.

"Earlier on, I was trying to convince people that climate change was happening," National Climate Adviser McCarthy said about the nearly half-century she has spent working on climate and environmental issues.[13] "I don't have to do that anymore."

Biden picked McCarthy to be the nation's first national climate adviser in part because of her ability to talk about climate change in ways that hit home with everyday Americans. While running the Natural Resources Defense Council, she pushed the nonprofit and its spokespeople to talk less about the problem of climate change and more about the solutions and opportunities that came with doing something about it. It's a narrative she stuck with after joining the Biden administration.

McCarthy doesn't speak about pollution in terms of particulate matter or parts per million, she talks about children in poor communities who are struggling with asthma. She doesn't talk about climate change in terms of disasters and woe to come in the future; she speaks of opportunities that come with clean energy, electric vehicles, and energy efficiency advancements today. And she doesn't talk about the Biden administration's plans in terms of sacrifice but rather in terms of shared prosperity.

"This is not about sacrifice, this is about investments," she told me. "This is about how we get out of the challenges we've been facing in the last couple of years. It's about how we regrow America. How we invest in ourselves. It's about winning the twenty-first century. We're now in a competition with many other countries for who's going to win in the clean-energy world. It has to be us."

To see how Biden's climate and clean-energy agenda will drive jobs and transform our economy while also combatting climate change, you have to look beyond the Beltway. Never before has the federal government supported clean energy and clean vehicles like it did beginning with the Biden administration. Not during the Carter administration, not during the Obama administration. Not any time before 2021.

But state governments have.

In 2004, Colorado became the first state to enact a renewable portfolio standard by ballot initiative.[14] The standard required the state's sixty or so utilities to get just 3 percent of their energy from renewable sources by 2007 and 10 percent by 2015. Just three years after the Colorado standard took effect, Vestas Wind Energy opened a thirteen-million-square-foot turbine tower plant and American manufacturing headquarters in Pueblo, in the state's eastern plains.[15] Shortly after that, it opened a wind turbine blade manufacturing plant in Windsor, about two hours away, as well as another blade and nacelle factory in nearby Brighton.[16] By 2020, Denmark-based Vestas had created about four thousand new jobs in Colorado and helped transform the state's economy from one dependent on oil and coal to one built on clean energy. Fifteen years after Colorado voters passed its renewable policy, about a third of the state's energy came from renewable sources. More than 60,000 Coloradans worked in clean energy. And the skies, slopes, rivers, and outdoors that Coloradans love so much were cleaner and better than they were a few decades earlier, bolstering Colorado's huge outdoor industry and the state's overall quality of life. In May 2019, Colorado Gov. Jared Polis signed into law legislation outlining the roadmap for his state to get 100 percent of its energy from renewable sources by 2040.[17] Doing so, Polis remarked, would create thousands more jobs and attract billions more in investments to the state. The proof was in the jobs, economic growth, and better environment that came with the state's earlier leadership on clean energy.

Colorado's renewable standard was the first enacted by referendum, but it wasn't the first. More than two decades earlier, Iowa became the first state to adopt a renewable portfolio standard by regulation.[18] Like Colorado, it too benefited greatly. Before being supplanted by Texas, Iowa produced more wind energy than any other state and led the country in wind energy jobs and investments. In 2015, I and a group of Iowa businesspeople traveled to Des Moines to meet with then-Gov. Terry Branstad. Branstad was first elected governor in 1983, the same year Iowa launched its nation-leading renewable energy standard. He would go on to be the longest-serving governor in American history. Branstad, a staunch Republican who was later ambassador to China under President Trump, didn't like to talk about climate change. His blood boiled and his face turned red with anger when the topic of federal environmental regulations came up. But when it came to wind energy and the twenty-eight thousand jobs created in Iowa by clean-energy companies, well, that made the paunchy governor happy and produced a smile under his bushy mustache. One of the things Branstad liked most about flying home to Iowa at night, he told us, was seeing the flashing red lights atop the five thousand or so Iowa wind turbines that produce energy used throughout the Midwest. "When I see those little lights flashing, I think about all the money coming into Iowa," he said with a sly grin. Branstad's successor, Gov. Kim Reynolds, also understands the economic importance of expanding clean energy. In 2017, after E2 released its annual Clean Jobs Midwest report, she issued a statement nothing that the report "highlights what we know to be true in Iowa: Renewable energy creates jobs and powers our economy." Iowa's wind industry also made a convert out of another powerful Republican: Iowa Senator Chuck Grassley. In 1992, seeing (like Branstad) the growth and ongoing potential for wind energy in his state, Grassley authored the legislation authorizing the nation's wind energy production tax credit. In 2020, when wind energy became the biggest source of electricity in Iowa, he held up his state as an example of how smart clean-energy policies can transform an economy.

"The proof is in the pudding," said Grassley, who sometimes calls himself the grandfather of wind energy in America.[19] Wind energy, he said, "creates thousands of jobs, supports economic development, boosts tax receipts, attracts investment in our state, and puts extra money in farmers' pockets. It delivers affordable energy for consumers that's also safe for the environment and helps build American energy independence."

Today, thirty states and the District of Columbia have followed Colorado and Iowa and implemented their own renewable portfolio standards, requiring utilities to get anywhere from 2 percent to 100 percent of their energy from renewables or carbon-free sources.[20] The Biden administration's plans for America to get 100 percent of its energy from clean, carbon-free sources by 2035 would essentially codify what the states have been doing, and up the ante for many of them.

As with renewable energy, states were way ahead of the federal government on energy efficiency. Since the 1970s, California has had some of the most stringent and effective energy efficiency policies in the world, requiring manufacturers to improve the efficiency of everything from refrigerators and computers to buildings and power plants if they want to do business in the state that is the fifth-biggest economy on the planet.[21] The state's per-capita electricity consumption has remained nearly flat the past forty years, while in the other forty-nine states, per-capita energy use has on average increased more than 50 percent. California has the fourth-lowest energy consumption per capita, according to Department of Energy data, and among states with more than five million people it is second only to New York in the lowest energy consumption per person. As a result, the state has saved an estimated $65 billion through energy efficiency since the 1970s. Along with savings, energy efficiency has created jobs, and lots of them. As of 2020, more than 53,000 energy efficiency–related businesses across the state employed 284,000 Californians, more energy efficiency workers than any other state and as many people who live in Toledo, Ohio or Montgomery, Alabama.[22]

Some of the same fundamental energy efficiency policies that California pioneered, including home efficiency standards, weatherization programs, and building electrification, will now be emulated elsewhere around the country through the Biden administration's policies.

Few legislators understand how smart climate and clean-energy policy can flow from states to the federal government and back out again like California's Fran Pavley. During her fourteen years in the California legislature, she authored the first legislation in the world to regulate carbon dioxide and other greenhouse gas emissions from motor vehicles. The legislation that came to be known as the Pavley Bill was later used by the Obama administration as a model for federal automobile emissions standards.[23] Pavley, a Democrat, also authored California's landmark Global Warming Solutions Act of 2006, the legislation signed by then-Gov. Arnold Schwarzenegger, a Republican, that set some of the first greenhouse gas emissions standards in the country and also became a model for the federal government, other states, and even other countries.

"The role of states is still incredibly important," Pavley told me.[24] "That's where you can reach people in messaging" in ways you can't at the national level. Besides, Pavley added, the federal government hasn't exactly been very successful passing climate legislation of any kind.

Twenty years after Pavley and her state led the country in setting greenhouse gas emissions limits for cars, California once again led the country in cleaning up the other big part of the transportation system—big rigs. In June 2020, the California Air Resources Board unanimously adopted the world's first zero-emission truck requirement. The state's Advanced Clean Trucks rule requires between 40 and 75 percent of all medium- and heavy duty trucks sold in California to be zero-emission by 2035, depending on the size of the vehicle. Just as Pavley's passenger vehicle emissions standards were later adopted by numerous other states and then used as a model for federal car standards—and along the way led to development of hybrid and electric vehicles and more efficient technology in other cars—the California trucks rule is doing the same. In 2021, New York, New Jersey, Washington and Oregon adopted the California trucks rule, and in March 2022, the Biden administration announced the first of three new regulations to reduce greenhouse gas emissions from heavy-duty vehicles, following the model set by California and other states. It was the first time in 20 years that the federal Environmental Protection Agency took action to update pollution standards for big rigs and medium-duty trucks. In addition to cutting greenhouse gas emissions and cleaning up pollution, switching from diesel-belching big rigs to electric and other zero-emission trucks is driving another huge shift in the vehicles industry—and our economy. In California alone, NRDC estimates the shift will save about $6 billion in fuel costs and create nearly 7,500 new jobs to build new electric 18 wheelers and delivery trucks, install charging stations at truck stops and otherwise clean up the industry. Nationally, shifting more of the nation's 12.2 million trucks that consume 46 billion gallons of fuel each year to electric or other zero-emission fuel sources is a potential game-changer for the trucking businesses, fossil fuel industry and the entire economy.

U.S. Energy Secretary Jennifer Granholm has seen it both ways. Prior to joining the Biden administration, she served two terms as governor of Michigan.

"It's an old cliché to say that the states are the laboratories of democracy, but it is totally true," Granholm told me during a webinar sponsored by E2 and NRDC.[25] "The states obviously have the ability to move faster and act bolder than the federal government is often capable of, as we have seen. And when that policy window finally opens in Washington, we're often starting out with what worked at the state level as a foundation for legislation that's going to make sense for the entire country."

Importantly, states continue to lead on climate and clean-energy policy, just as they did back when Fran Pavley was in the California legislature, just as they did during the Trump administration, and just as they continued to do even amid all the Biden administration's federal climate and clean-energy initiatives.

In October 2021, for instance, North Carolina Gov. Roy Cooper signed into law a sweeping energy bill designed to reduce emissions from the state's power plants by 70 percent by 2030 and 100 percent by 2050.[26] The law is expected to dramatically boost the deployment of solar and offshore wind in the state and add North Carolina to a list of sixteen other states that have enacted legislation establishing greenhouse gas–reduction requirements.

In September 2021, Illinois Gov. J. B. Pritzker signed the Climate and Equitable Jobs Act, which set his state on a path to get 100 percent of its power from clean energy by 2050, with a focus of creating clean-energy jobs for people and communities of color.[27] A month later, Pritzker signed Illinois' Reimagining Electric Vehicles in Illinois (REV) Act, which provides tax breaks and other incentives for electric vehicle manufacturers and suppliers to build in his state.

"We've already taken so many incoming calls already . . . from all the major car companies and all the major entrepreneurial start-up companies around the country that are looking to where they are going to put their factories, where they are going to build their new vehicles or their assembly plants or their battery factories or their supplier vendor companies," Gov. Pritzker told me during a webinar sponsored by E2 and NRDC.

Not to be outdone, New York Gov. Kathy Hochul in January 2022 announced plans for her state to become the first in the country to ban all natural gas hook-ups from new buildings (expanding on a similar ban passed by New York City in 2021). She also announced other sweeping climate policies, including plans to invest $500 million to expand offshore wind and establish a dedicated "green electrification fund" to electrify low-income housing and train the workers needed to do it.

To be sure, clean-energy and clean-transportation businesses are eager, ready, and waiting for the climate and clean-energy legislation being rolled out in states and nationally. In June 2021, emboldened by the Biden administration's announcement to get 100 percent of the nation's energy from renewables, solar panel manufacturer First Solar announced plans for a third, $680-million factory near Toledo, Ohio, that will create five hundred new jobs, in addition to the 1,600 workers it already employs in the city.[28] "We have said that we stand ready to support President Biden's goal to transition America to a clean, energy-secure future, and our decision to more than double our US manufacturing capacity with this new facility is First Solar making good on that commitment," company CEO Mark Widmar said in a statement. Later that same month, while Congress was debating the Build Back Better Act, General Motors announced plans to invest $35 billion in electric and autonomous vehicles by 2025, including building a new factory to manufacture batteries for its vehicles and—through a joint venture with Wabtec Corp.—also for locomotives.[29] If you think electric powered long-haul trains are far off in the future, think again. In January, Union Pacific Railroad agreed to spend $100 million to buy 20 battery powered locomotives from Wabtech and another battery maker, Progress Rail. Plans call for the new locomotives to hit the rails in 2023. In September 2021, Ford Motor Co. announced the single-biggest investment in its history: $11.4 billion to build four new electric vehicle factories. And in October 2021, Spanish renewable energy giant Siemens Gamesa announced it would build America's first offshore wind blade factory, creating an estimated 260 new full-time jobs.

Just as significant as the huge investments in clean energy and clean vehicles is where they are happening. GM isn't building its new factories in its hometown of Detroit or even in Michigan, but in Tennessee and Ohio. Similarly, Ford isn't building in Motown either. Its plans include a new $5.6-billion megacampus in Stanton, Tennessee, that it will call BlueOval City, after the automaker's iconic logo, creating six thousand jobs there. That's about ten times the number of people who live in the town.[30] In December

2021, buoyed by the strong debut of its first electric trucks, Irvine, Calif.–based Rivian announced it would expand its manufacturing opreations with a second factory near the little town of Rutledge, Georgia, about an hour east of Atlanta. Rivian eventually expects to employ 7,500 workers there—about 15 times the number of people who live in Rutledge. And while First Solar picked Toledo, Ohio, for its expansion, wind energy company Siemens Gamesa selected an eight-acre site in Portsmouth, Virginia, for the nation's first offshore wind blade factory. The company was lured to the region after the state of Virginia and Dominion Energy announced plans for a 2.6-gigawatt offshore wind project. As Liz Burdock, CEO of the Business Network for Offshore Wind, said, the signals being sent by both the state and federal government are "a welcome sign of growing investor confidence in the U.S. market."[31]

Ford, GM, Rivian, Siemens, First Solar, and other companies are transforming America's economy and turning small towns into national hubs for clean energy and clean vehicles. It's not dissimilar from what the oil industry did for Texas, or what the tech industry did for Silicon Valley. Or perhaps more accurately, what Tesla did when it opened its seven-thousand-employee Gigafactory in little Sparks, Nevada, transforming the dusty town and surrounding Storey County into the area that in 2018 had the highest density of clean-energy jobs per capita in the country.[32]

The boom in electric vehicles is transforming the Southeast in particular, not unlike how textiles, tobacco, banking and biotech transformed the region previously. Along with attracting major operations from Ford, GM, Rivian and other electric vehicle makers and suppliers, the Southeast is now beginning to attract follow-on businesses—including some from other continents. In February 2022, for instance, Australian car charger manufacturer Tritium announced it would open its first U.S. manufacturing plant in Lebanon, Tenn., not far from Ford's Blue Oval City and other EV car companies. The factory, Tritium's biggest ever, is expected to employ 500 workers. At a White House ceremony with Tritium president and CEO Jane Hunter, President Biden made note of the ripple effect of the auto industry's transition to electric vehicles. "They'll use American parts, American iron, American steel and . . . and (Tritium's chargers) will be installed up and down highways and corridors . . . all across the country," Biden said. "The benefits are going to ripple in thousands of miles in every direction. These jobs will multiply in steel mills, small parts suppliers, constructions sites all over the country for years to come."

Just as auto manufacturers are shifting to 100 percent electric vehicles in response to federal and state policies and declining technology prices, electric utilities across the country also are overwhelmingly shifting to clean energy.

So are big users of electricity, who suddenly realize that they can produce their juice themselves. That in turn is creating opportunities not just for big companies like First Solar and Siemens Gamesa but also for small businesses that now can be players in a more open market once limited only to big investor-owned utilities. That includes companies like Volt Energy Utility. The small, minority-owned-and-operated solar company got its start in Winston-Salem, North Carolina, in 2009, just as Recovery Act funding and North Carolina's renewable energy standard were pushing utilities in the Tar Heel State to use more solar. Volt's early projects included major solar installations for Wake Forest University and the U.S. Army base at Fort Bragg, North Carolina. Since then, it has relocated to Washington, D.C., and has developed solar projects for the corporate headquarters of The Cheesecake Factory and a Subaru dealer in Colorado, as well as for the District of Columbia and the D.C.-area utility Pepco.

Volt CEO and cofounder Gilbert Campbell said he's personally energized by lawmakers' attention to equity and improving opportunities for minority-owned companies like Volt as they roll out new clean-energy policies and spending.[33] "Right now, we have the chance of a lifetime to expand clean energy and the jobs, savings, and other economic benefits that come with it to every part of America, including to communities of color and businesses run by people of color," Campbell said. "Climate change is showing us every day we can't wait any longer." Small businesses like Campbell's will be among the biggest beneficiaries of America's economic transition to clean energy. According to E2 research, about two-thirds of all clean-energy businesses have fewer than twenty employees. Nearly 90 percent of clean-energy companies have fewer than one hundred employees.[34]

Similarly, minority-owned businesses like Campbell's could also see a boost because of the Biden administration's whole-of-government focus on equity with its climate and clean-energy plans. Under the administration's Justice 40 initiative, at least 40 percent of the overall benefits of federal investments in climate and clean energy must be directed at disadvantaged communities and communities of color. That will help ensure more funding for weatherization and renewable energy projects in inner cities, solar and wind developments in rural areas, and workforce training investments in fossil fuel communities.

"Part of this is about where we invest money," said McCarthy, the national climate advisor. "We have a real priority to expend money into the communities that need it the most." The Justice 40 initiative, she said, is designed to make sure money for climate resilience goes to parts of the country that are most vulnerable to climate change, and that money for clean-energy investments goes to low- and moderate-income communities and communities of

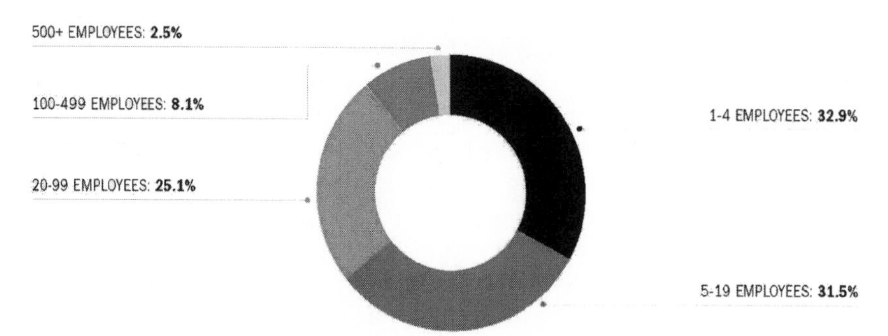

Figure 10.2. Small businesses with fewer than 100 employees employ about 90 percent of all clean-energy workers in America. *Source*: E2, "Clean Jobs America, 2021."

color. In other words, nearly half of the government's investments in clean energy and climate resilience are supposed to be directed to the communities that are bearing the brunt of climate change and pollution, but so far have seen the least benefits of clean energy.

Critics say the plans sound good on paper but that the Biden administration needs to move faster. And since Congress, not the executive branch, ultimately controls the government's purse strings, how federal dollars are doled out isn't completely up to the administration. To be sure, the Biden administration's financial leash would get much tighter if Republicans take over Congress.

The Biden administration is relying on its equity focus to try and bring good-paying jobs that come with benefits—many of them union jobs—to parts of the country that need them the most. According to analysis by BW Research for E2, the median hourly wage for clean-energy workers was $23.89 per hour in 2019, about 25 percent more than the national median of $19.14 per hour.[35] Jobs in wind energy pay the best, with a median wage of nearly $26 per hour, while energy efficiency jobs paid a median of about $24.44. Furthermore, clean-energy jobs were more likely to come with health care and retirement benefits than jobs across the rest of the private sector.

Better wages and benefits are common across the gamut of jobs in clean-energy businesses. For instance, the median pay for all welders in 2019 was about $17.31 per hour, but if you were a welder working in renewable energy that year, you made closer to $21 per hour, according to BW Research's analysis of U.S. Bureau of Labor Statistics (BLS) data. Most electricians could expect a median wage of about $27 per hour, but if you were an electrician working in grid modernization, renewable energy, or energy efficiency in 2019, you could expect an additional $2 or so per hour. According to BW

WAGES AND UNIONIZATION RATES BY INDUSTRY, 2019

	2019 Median Hourly Wages[8]	% Above or Below the National Median	Unionization Rates[9]	Nationwide Employment, 2019[10]
U.S. TOTAL	$19.14	-	6%	149,133,921
CLEAN ENERGY INDUSTRIES OVERALL	$23.89	25%	9%	3,355,419
Renewable Energy Generation*	$23.44	22%	6%	522,811
Solar	$24.48	28%	4%	345,393
Wind	$25.95	36%	6%	114,774
Energy Efficiency	$24.44	28%	10%	2,378,893
ENERGY STAR Appliances	$24.63	29%		142,272
Renewable Heating & Cooling	$24.91	30%		129,998
Efficient Lighting	$24.21	26%		380,299
Traditional & High-Efficiency HVAC	$24.43	28%		1,034,666
Grid Modernization & Storage	$25.07	31%	12%	147,644
Grid Modernization	$25.40	33%		67,945
Storage	$24.82	30%		79,699
Clean Fuels (Advanced Biofuels)	$19.55	2%	6%	39,704
Clean Vehicles	$22.20	16%	9%	266,368
Other Industries				
Retail Trade	$13.16	-31%	4%	15,613,400
Accommodation	$11.64	-39%	7%	2,084,600
Food Services & Drinking Places	$11.48	-40%	1%	12,026,200
Arts, Entertainment, & Recreation	$13.88	-27%	7%	2,415,400

* Includes low-impact hydropower, combined heat and power, bioenergy, and geothermal in addition to wind and solar.

Figure 10.3. Clean-energy jobs pay about 25 percent better than the national median wage. *Source*: E2, ACORE, CELI, BE Research, "Clean Jobs, Better Jobs (2021)."

Research's analysis, about 9 percent of clean-energy jobs in 2019 were unionized, compared with about 6 percent of jobs in the overall private sector. To be sure, some clean-energy sectors—solar and wind in particular, along with EV upstarts like Tesla—eschew union labor, creating a perplexing problem for both unions and the Biden administration. But other clean-energy sectors, such as electric grid modernization (unionization rate is about 12 percent) and manufacturing jobs making Energy Star appliances, high-efficiency HVAC systems, and heavy equipment (energy efficiency's unionization rate is about 10 percent) help offset the gap. As traditional auto companies produce more electric vehicles, more of the jobs needed to manufacture them are likely to be unionized, too.

While the Biden administration wants to bring more of those good-paying, union-friendly clean-energy jobs to every part of the country, it really is trying most just to build on what the private-sector is already doing.

"There's a change in the world that's already happening," McCarthy told me. "What we can do is accelerate it. This is the way the United States wins in the twenty-first century."

Chapter Eleven

Following the Money

While government policies and leadership from Washington can help accelerate change, there's another place that can accelerate change much faster: Silicon Valley.

In 2003, as a national technology reporter for a chain of newspapers, I visited the Mountain View, California campus of Google to meet with cofounder Sergey Brin. At the time, Google was still a private company, though there was widespread speculation that it would launch an initial public offering soon. The moment I pulled into the company parking lot, I got a taste that Google wasn't a typical company. Covering many of the parking spaces were canopies made from solar panels, something that's commonplace today but back then was pretty unusual. Even more unusual were the thick power cords hanging down from the panels over nearly every parking space, something that didn't make sense until Brin and team later explained it to me. At the time, electric vehicles were even more uncommon than solar parking lot canopies (the first Tesla wouldn't hit the streets for another five years). But Google knew EVs were coming someday soon, and it wanted to be ready. Google also wanted employees and other visitors to think about the possibilities that could come with solar-powered parking lots and cars that you could plug in to refuel.

Two of the forward-thinking people responsible for Google's early solar deployment were Chris Sacca, who as the company's corporate counsel and later head of special initiatives was involved in Google's energy purchase agreements, and Andrew Beebe, who was chief commercial officer at solar company Suntech, which helped Google go solar.

"There really wasn't any corporate interest until those guys stepped up and said, 'Please build solar arrays all over our campus,' Beebe recalled during a Greenbiz Verge clean-tech conference in October 2021.[1] But (Google executives) also said, "Set it up so we can have Walmart and Cisco and Microsoft and all of our competitors come over and see what we have done." They obviously had a hugely catalytic role in making all this happen."

Both Beebe and Sacca would go on to become successful venture capitalists, Beebe with Obvious Ventures, the firm that helped launch companies such as Medium, Beyond Meat, and electric bus maker Proterra, and Sacca with his firm called Lowercase Capital, which funded companies such as Twitter, Uber, and Instagram. For about three years, Sacca also was a "guest shark" on the ABC television show *Shark Tank*, where budding entrepreneurs bid for the favor—and the funding—of millionaire investors. But it didn't take long before Sacca was feeling unfulfilled by funding kitchen gadget start-ups on *Shark Tank* or electronic-gaming companies back in Silicon Valley. He, like Beebe, turned his attention almost fully toward clean-energy and climate-related investments.

Sacca and Beebe represent one of the hottest corners of the venture capital business in the 2020s: Climate Tech. Some of the companies that investors like them are backing today will likely become the Googles of tomorrow. Only instead of changing the way we search for stuff on the Internet, climate-tech companies will change the way we source and store our energy, grow our food, and move from point A to point B, whether on land, water, or air. In doing so, they'll not only transform our economy, but help save the planet.

In 2021, investments in climate-tech companies hit more $31 billion, according to deal tracking firm PitchBook.[2] That was 30 percent more than in 2020 and more than two-and-a-half times what it was in 2019. Those big numbers will likely only get bigger as federal, state, and international clean-climate and clean-energy policies are implemented. Quite simply, government policies and funding help reassure venture capitalists and other private investors to put more of their money at risk.

Climate-tech and clean-tech investing is no longer just about solar or wind or even batteries anymore. Those businesses now attract plenty of mainstream investors. They're almost like investing in restaurants or real estate—they're too passe for venture capitalists who are more interested in finding more disruptive technologies that can scale quickly and create big returns.

"What we look at every day are energy innovations that are just insane, some of which are doing things that Einstein declared literally would not be possible," Sacca said at the Verge conference.[3] "We see stuff happening in synthetic biology, for instance, that's just nuts."

Amid the hellish fires in the West, back-to-back hurricanes in the East, and scientists everywhere warning that things were only going to get worse, Sacca in August 2021 stepped away from Lowercase Capital, quit *Shark Tank*, and with wife Crystal turned his attention specifically toward figuring out how to fund and support companies trying to do more to address climate change. The couple launched a new investment fund called Lowercarbon Capital. In a matter of days, they raised more than $800 million that Lowercarbon Capital could deploy to try to "unfuck the planet," in Sacca's terms. The fund was so popular, Sacca wrote on Lowercarbon Capital's blog, that it had to turn investors away.[4] "It turns out that raising for a climate fund in the context of an unprecedented heatwave and from behind the thick clouds of fire smoke probably didn't hurt," he wrote.

Since then, Lowercarbon has invested in companies that capture carbon dioxide and turn it into consumer products, reduce carbon emissions from livestock and fertilizers on the farm, and mine materials that are key to batteries and storage in ways that don't destroy the environment.[5] One such company is Twelve, a Bay Area start-up that "upcycles" carbon dioxide and carbon monoxide captured from industrial emissions and turns it into everything from jet fuel to sunglasses lenses, replacing fossil fuels and plastic. Another company Sacca was particularly excited about in 2021 was Lilac Solutions, which has raised $150 million to commercialize its lithium-mining technology. Lilac claims it can produce the essential element for batteries ten thousand times faster than conventional methods, using 90 percent less land and water. Lowercarbon Capital has also made numerous major investments in companies at the intersection of agriculture and climate, including start-up Formo, which is following the Beyond Meat and Impossible Burger model to make fine European cheeses that don't require dairy or cows; Entocycle, which has figured out how to speed up the gestation period for black soldier fly larvae which happen to be some of the world's fastest converters of food waste to protein; and Nitricity, which uses solar-powered modules placed around farms to literally make fertilizer out of thin air by converting and processing nitrates found in the atmosphere.

If garbage-eating fly larvae and fine cheeses bioengineered in a sterile laboratory don't sound like appealing business models, think again. According to research group Climate Tech VC, food-and-water-related climate tech was the biggest sector for climate venture funding in 2021, followed by mobility, consumer goods, and clean energy.[6] Tech investors' take on food and agriculture is yielding new high-tech twists in one of the world's oldest and most established economic sectors. Seattle-based clean-agriculture start-up Nori, for instance, got its start in 2017 when its cofounders entered a hack-a-thon contest for coders to figure out new ways to use blockchain technology for social good. Far from the nearest farm, what they came up with was a way to use blockchain technology to monitor and track low-carbon agriculture practices and then monetize that by selling farm-based carbon-removal offsets. In doing so, Nori is incentivizing farmers to use more climate-friendly agriculture practices that don't just reduce carbon emissions but actually increase the ability of soil and crops to store carbon, while also creating a new marketplace for carbon removal and trading. In 2020, Nori raised more than $5 million in seed funding to launch its platform. "We call it climate-smart agriculture—thinking of carbon removal like a crop," Christophe Jospe, a Nori cofounder, told E2.[7]

Climate-tech investments are starting to revolutionize the transportation sector too, well beyond electric cars and trucks. Consider, for example, high-flying Heart Aerospace. For years, the Swedish company has been working on a nineteen-seat, battery-powered airplane that it had hoped to sell to a major aircraft company like Boeing or Airbus. The big companies all told little Heart there just wasn't a market for small electric planes, even if the company could figure out the technology to make them fly. Yet as countries in Europe and elsewhere levy carbon taxes on jet fuel and airlines learn from the tectonic shifts happening in the automotive industry, Heart's electric planes are looking increasingly more attractive. In July 2021, Heart raised $35 million in funding from investors including Bill Gates's Breakthrough Energy Ventures, EQT Ventures, and Sacca's Lowercarbon Capital.[8] Even more significantly, other investors included United Airlines and Mesa Air Group, which, as part of their funding agreements, also placed orders for the first two hundred of Heart's planes, to be delivered in 2026. Heart claims its first-generation planes will be able to fly about 250 miles between charges. That means United and Mesa could use them to fly short but profitable hops between cities like San Diego and Los Angeles, or Chicago and Indianapolis, or even New York and Washington without burning a drop of fossil fuel. And just like electric cars, electric planes are cheaper to operate than fossil-fuel-powered planes. Heart claims its electric motors are twenty times less expensive than similar-sized turboprops, and that its maintenance costs are more than one hundred times lower. That makes them even more attractive to companies like United and Mesa Air.

Heart Aerospace isn't the only next-generation airplane company getting attention—and funding—from high-profile backers. ZeroAvia is a company with offices in London and California that is developing a hydrogen-powered aircraft that can fly as many as one hundred passengers for longer-distance flights. It plans for its first flight to take off in 2024 between London and Rotterdam. In March 2021, the company raised $24 million in third-round funding, increasing the amount of money it has raised to about $70 million from funders including Gates's Breakthrough Energy and Amazon founder Jeff Bezos's Amazon Climate Pledge Fund, as well as British Airways and the investment arm of Royal Dutch Shell.[9]

The participation of fossil fuel giant Shell as well as traditional airlines like British Airways, United, and Mesa shows how Silicon Valley–style investors are pulling old-school companies and their money into climate action not necessarily because they want to save the planet but because they want to preserve their businesses and keep making money in a fast-changing economy that's becoming cleaner. Nine months after its investment in ZeroAvia, Shell announced plans to invest in a major new biofuels plant at its Rotterdam Energy and Chemicals Park. As the world's biggest supplier of aviation fuel, Shell announced at the time that at least 10 percent of its aviation fuels will be sustainably sourced by 2030—a good start, even if it is probably too little, too late, given the state of the climate crisis. Still, for conscientious investors like Sacca, seeing big old-school companies try to get it right on climate—even if they're driven only by profits—is a good sign.

"We see companies getting there by guilt or shame, which I don't think actually scales up very well," Sacca said at the Verge conference. "Or they get there by taking pride in what they're doing . . . or they get there by sheer greed—which I think is the most positive motivating factor, the most effective factor in changing systems on this planet."

As Sacca and Lowercarbon Capital like to say, "fixing the planet is just good business. Shame and guilt won't get us there, markets will."

Of course, venture capitalists have bet on clean tech before, and many failed miserably. From 2006 to 2011, venture capital firms sunk more than $25 billion into clean-tech start-ups, and they lost about half of it, according to a study from MIT.[10] Money flowed into thin-film solar companies that were going to turn everything with a surface into energy-generating solar panels before they flopped. More money went into now-bankrupt companies that wanted to roll out nationwide chains of battery-swapping stations for electric vehicle drivers. Other money went into exotic energy storage technologies that were supposed to make batteries obsolete, but of course, they didn't. "We conclude that the VC model is broken for the clean-tech sector," MIT researchers wrote in 2016.

What's different today is that technology, public funding, and the public need for climate solutions has caught up with climate tech. Just as the world wasn't quite ready for solar panels in Jimmy Carter's days, it wasn't quite ready a decade ago for the types of climate-tech solutions we're seeing today. And with the economic costs of climate change rising every day, new innovation and technology is needed more than ever. The timing is right.

Eric Berman has been a clean-tech investor since 2006, when he joined a group of other Seattle-area investors in E8, one of the nation's first clean-tech-angel investor networks. Since 2006, E8 members have invested more than $40 million in one hundred companies across the United States and Canada that focus on climate tech, clean energy, and sustainability.

"What attracts me to this is that I believe saving the planet is not only not costly, it's actually profitable," said Berman.[11] He worked at Microsoft for eight years and another six as vice president at Microsoft spin-off Expedia before leaving the company in 2005. "We're making bets that we believe are in fact profitable. And if we're wrong, the world as we know it is largely screwed."

Berman told me he's more optimistic about climate and climate tech than he has been in a long time. First, he said, that's because the "slow money, the risk-averse money" is now steadily flowing to clean energy. Utilities and big users of electricity are now buying as much renewable energy as they can get because it's now the cheapest power available in most parts of the country. They're willing to sign long-term agreements to get it, too. According to Berman, that sends a solid market signal and creates a stable foundation on which Silicon Valley innovators can build upon with even more disruptive technologies, just like so many start-up companies did on the foundation of the Internet or mobile communications networks in the past. Meanwhile, clean-energy policies at the federal, state, and international levels are finally starting to move in the right direction and in the right ways to really accelerate change in the marketplace and throughout the economy. "There are three things that affect change: Advocacy, policy, and markets," Berman said. And now, he added, all three things are lining up.

What's also encouraging for investors like Berman is that those same types of market dynamics are now beginning to ripple through the auto industry. Electric vehicles keep getting cheaper and cheaper and clean-vehicle policies at both the state and federal level keep getting better and better. As a result, the markets are now reacting, making it clear that EVs will soon become the vehicle of choice for businesses, consumers, and government agencies alike, just as renewable energy is now the electricity of choice for many. Big buyers of vehicles—whether delivery companies like Amazon, UPS, or FedEx, or rental companies like Hertz and Avis—are now buying clean vehicles in big numbers, just like utilities and big energy users did with clean energy. In December 2021, the world's biggest buyer of goods—the federal government—accelerated that shift when President Biden ordered government agencies to purchase only zero-emission vehicles and zero-emission electricity going forward.[12]

Another investor who understands policy and markets, and who is firmly focused on climate tech, is Tom Steyer, the billionaire and former presidential candidate who made climate action the foundation of his 2020 run for the White House. In September 2021, Steyer and longtime business partner Katie

Hall launched Galvanize Climate Solutions, a new investment platform that will raise money and invest solely in companies and organizations working to reduce carbon emissions around the world. In addition to helping game-changing companies get going by giving them the capital to do it, Galvanize also plans to leverage Steyer's and other principals' political connections and policy prowess to help make it easier for businesses to play a bigger role in advancing smart climate and clean-energy legislation. Early backers of the fund include a veritable who's who of Silicon Valley, such as Salesforce CEO Marc Benioff and an investment firm controlled by philanthropist Laurene Powell Jobs, the widow of Apple founder Steve Jobs.[13]

One of the things that makes Steyer confident about investing in climate technology now is that businesses, government, and the public at large in America has finally moved beyond discussing whether climate change is real and now is moving on to doing something about it.

"The conversation about whether we need to do this . . . that conversation in a very real way in the business world is over," Steyer said at a forum at the COP26 climate summit in Glasgow, Scotland.[14] As a result, he added, "We're seeing the kind of explosion in this area that we saw in telephony and IT around the end of the last century. We're seeing now an Internet kind of revolution."

Steyer and other wizards of Silicon Valley are jumping feet first into climate-tech solutions based on what they learned from—and profited from—in the world's last great economic transition, the technology transition. Nobody wants to be the investor that sat out an opportunity to invest in the climate-tech version of Google or Tesla. Everybody wants to be a part of the solution for saving the planet. And now, they're more confident they can make money doing it.

"I think what we are seeing is that it no longer has to be exclusive to have a massively positive impact on the planet and make a ton of money," Sacca said.[15] "It wasn't always that way. But the economics have shifted."

The economics have shifted, too, in a way that tech investors understand. Silicon Valley companies beginning with Fairchild Semiconductor and Hewlett-Packard, and later Apple, Google, Facebook, and thousands of others, revolutionized the way the world communicates, buys goods, and searches for information—and forever changed our economy. But those companies couldn't have changed the world if not for the early government funding of research and development that was essential to the creation of the integrated circuit, batteries, and the Internet itself. The same thing is now happening with climate tech. Federal climate and clean energy policy, the international agreements forged at climate summits in Paris in 2015 and Glasgow in 2021, and state-level clean-energy policies represent in some ways the last big government push of the climate action sled to the top of the hill. Now, private investors and companies are jumping on board to steer it through the downhill slope, the place where climate action combined with business will accelerate the fastest and make the most difference in both transforming our economy and saving our planet.

"We absolutely need government leadership—value-driven, justice-driven government leadership—to set up a framework for what people are allowed to do in the private sector," Steyer said.[16] "But when you think about where innovation in this world has happened in the twenty-first century, the twentieth century, the nineteenth century . . . where has the innovation and true advances come from? They've come out of companies."

Chapter Twelve

Politics, Jobs, and Saving the Planet

For someone at the center of the nation's environmental policy making for more than a decade, there aren't a lot of environmental policies Senator John Barrasso likes.

He derided the support of wind, solar, and other clean-energy investments under the 2009 American Recovery and Reinvestment Act as "boondoggles."[1] He introduced legislation to prevent the Environmental Protection Agency from limiting carbon dioxide emissions.[2] He has voted repeatedly to pull the plug on electric cars, introducing numerous bills that would discontinue tax credits for EVs and instead levy new taxes and fees on EV buyers.[3,4]

Yet while Barrasso was in Washington bashing clean energy and bolstering the oil and gas industry, clean energy companies were growing dramatically in his home state, as oil and gas companies were shutting down and pulling out. Nothing illustrates this better than what happened in August 2020, when workers in Wyoming's Sweetwater County laid down what at the time was the only remaining working petroleum gas rig in the entire state, according to the *Casper Star-Tribune*.[5] With the exception of a few days in June 2020, the working rig count in Wyoming had not previously hit zero since 1884, back before Wyoming was even a state. About the same time the Sweetwater rig was coming down, less than three hours away in Carbon County, Wyoming, hundreds of workers were on the job erecting a thousand wind turbines at what is expected to be the nation's biggest wind farm, the Chokecherry Sierra Madre development.[6] Within a year, PacifiCorp., the parent company of Rocky Mountain Power, would announce plans for six more major Wyoming wind-and-battery storage projects expected to generate thousands more clean-energy jobs, along with enough electricity to power 350,000 homes throughout the Mountain West.[7]

It's a story that's been developing over a decade or more. Between 2011 and 2021, the number of Wyomingites working in mining, petroleum, and logging operations declined by nearly thirteen thousand, according to Labor Department statistics. During that same period, the number of jobs in wind, solar, and energy efficiency in Wyoming increased to about eight thousand.[8] At the beginning of 2021, more people still worked in fossil fuels (about 18,000 workers) than in clean energy in the state, but that's changing. Within a few years, the number of Wyomingites who work in clean energy could outnumber those who work in fossil fuels, leaving politicians like Barrasso, who has received more than $1 million in donations from oil and gas companies, in a lurch.[9]

The dynamic playing out in Wyoming is playing out across other Republican states as well. As it does, it's forcing Americans—conservatives, liberals, and everyone in between—to change the way they view climate change and clean energy. Regardless of their political views, they can now see climate-change-related weather disasters with their own eyes and feel it in their own pocketbooks. They can also now see the economic benefits of climate action with every new solar or wind project and every new electric vehicle on the road.

California by far has more electric vehicle registrations than any other state, but no. 2 for EV registrations is red-state Florida, and no. 3 is even-redder Texas, according to Department of Energy data.[10] Look for that shift to continue as EV prices continue to decline and pickup truck manufacturers continue to sell more selling electric or hybrid models, including Ford and its best-selling F-150. Similarly, now that Walmart and IKEA sell solar panels; now that LED light bulbs dominate Home Depot's lighting section; and now that Energy Star appliances and Nest thermostats are everywhere, the benefits of clean energy are literally hitting home with Americans everywhere, regardless of location or political persuasion.

Some politicians are beginning to get the message. Take what happened in July 2020. Four months before election day, seven Republican senators banded together to push Senate Majority Leader Mitch McConnell to include federal support specifically for clean energy in the country's economic stimulus bill. These weren't just some left-leaning conservatives from California or New York. They included Republican flag-bearers Senators Lindsay Graham, Thom Tillis, and Richard Burr from the Carolinas, and oil industry boosters Cory Gardner of Colorado and Lisa Murkowski of Alaska. "As we focus on getting the country back to work, we must include an industry that had already been putting Americans to work faster, and in more places, than the overall economy before the COVID pandemic hit," the Republican senators wrote in their July 23, 2020, letter to McConnell. "Clean-energy job growth has outpaced the economy by 70 percent over the past five years. Further, this growth is truly nationwide, encompassing every state."[11]

To be sure, congressional Republicans will likely oppose anything and everything from President Biden or any other Democratic office holder as vociferously as they possibly can, especially if they regain control of Congress. But they do so not necessarily because they don't like or want clean energy or climate action. They do so because GOP leaders tell them they must not allow their political opponents to win at anything. "One-hundred percent of our focus is on stopping this new administration," Republican Senate Minority Leader McConnell said in May 2021.[12] It echoed his proclamation a decade earlier that the primary goal of Republicans in Congress was to make Barack Obama a one-term president. Clean energy, climate change, possibly saving the planet—those things just don't matter as much to McConnell and Republican leaders in Congress as regaining control of Congress and the White House for their party. The problem is, opposing clean energy and climate action also means opposing what the majority of Americans say they want. It also means opposing the economic benefits accruing in their own states.

Look at where the jobs and economic growth from clean energy is happening. Not surprisingly, staunchly Democratic California leads the country in clean-energy jobs by far, according to E2's Clean Jobs America research.[13] But red states Texas, Florida, Ohio, and North Carolina are all in the top ten. Democratic congressional districts in 2020 had slightly more clean-energy jobs than Republican districts, but that could change if Republicans take more seats in Congress after the 2022 midterm elections. In 2020, congressional districts held by Democrats were home to about 1.6 million clean-energy jobs, while about 1.3 million clean-energy workers lived in Republican districts. In 2021, Democrats Rep. Nancy Pelosi and Rep. Ro Khanna of California represented the congressional districts with the most clean-energy jobs. But no. 3 was the deep-red district held by Texas Republican Rep. Dan Crenshaw, the former U.S. Navy SEAL from the Houston area who has derided solar and wind as "silly solutions." Utah Republican Rep. John Curtis's district was no. 4 in the country for total clean-energy jobs. Curtis, who cofounded a Provo, Utah, shooting range equipment company before getting into politics, has embraced clean energy and climate action more than most Republicans. In June 2021, he cofounded the Conservative Climate Caucus, a group of more than sixty Republican members of Congress who acknowledge climate change and embrace free-market solutions to address it, although they maintain fossil fuels are somehow part of the solution.

It's not public protests or climate science that are convincing Republicans to get more engaged on climate. It's economics. Solar, wind, batteries, and electric vehicles are just getting cheaper, while oil, gas, and coal are just

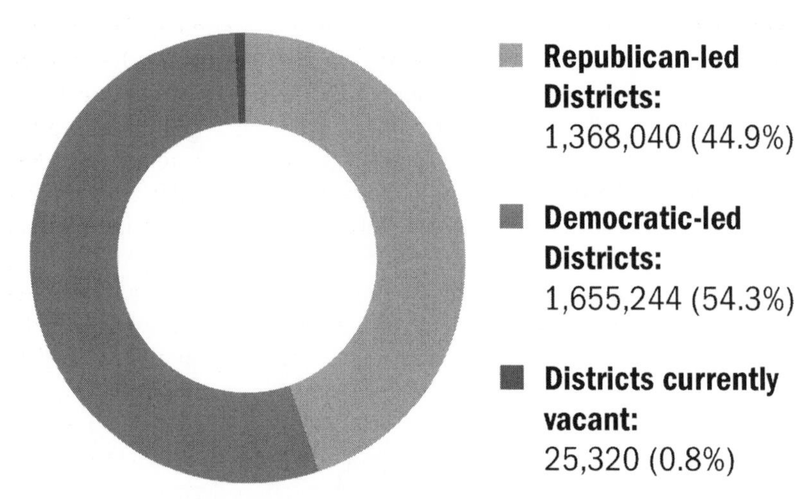

Figure 12.1. There are nearly as many clean-energy jobs in Republican congressional districts as there are in Democratic congressional districts. ***Source*****: E2,** ***Clean Jobs America—Congressional Districts (2021).***

getting more expensive. Climate disasters are striking every state, not just in Democratic states. And if Republican lawmakers want to attract the fastest-growing jobs in America to their home states, they realize they have to do more to embrace clean energy.

There's something else about climate change that should be concerning to politicians: Future voters. "If Republicans don't make it an issue, we will lose the upcoming generation of Republicans," Rep. Curtis told the *Salt Lake Tribune.*[14] "The upcoming generation will not be patient with us. This is a deal-breaker for them. They'll leave the Republican Party over this one issue." Polling bears that out. In a February 2020 survey of eighteen- to thirty-five-year-olds by the American Conservation Coalition (ACC), an organization of young conservatives for the environment, 77 percent of right-leaning young Americans and 90 percent of independents said climate change was an important issue to them.[15] About 90 percent also said the United States should invest more in clean energy, infrastructure, and natural ways to reduce emissions and combat climate change. "The tide is turning on climate change," said ACC president Benji Backer, who founded the group while in college at the University of Washington. "For young people, the issue is no longer about political affiliation . . . it's about results."[16]

Backer and other young conservatives realize they'll be the ones to deal with the existential and economic costs of climate change. But many conservatives also realize the economic opportunities that come with clean energy and climate action.

Mark Fleming has spent a lifetime at the intersection of Republican politics and business. He worked for former North Carolina Republican senator Lauch Faircloth and later, Republican U.S. Rep Patrick McHenry. In between, he ran the Wake Forest, N.C. Chamber of Commerce and the conservative free-market group NC FREE. It was in 2010, following a Republican sweep of the state legislature and governor's office in North Carolina, when Fleming clearly got the connection between the economy and climate change. At the time, he was representing his boss Rep. McHenry at a training session for newly elected Republicans that included a forty-five-minute session on climate change. Most of the session was dedicated to the environmental and health consequences of climate change, and the political framing of it, he recalled.

"And then at the very end, they talked about an emerging clean-energy economy in North Carolina, and the jobs it could provide," Fleming recounted. "I remembered thinking at the time that if they had led with the economic benefits and ended with the 'Oh by the way, this is helpful with climate,' they would've had everyone in the room ready to go with them."

Four years later, Fleming launched Conservatives for Clean Energy, which "proudly educates the public and decision-makers on the economic benefits of clean energy and advocates for continued investments across the Southeast," according to its mission statement.

Conservative politicians who support climate action and clean energy do so for several reasons, Fleming told me.[17] There's the desire to support free-market principles and struggling rural communities: "Some of it's about the big utility versus the little guy . . . and saving the family farm through things like leases for solar." Some of it is about reputation and religion: "They don't want to look like the Earth-is-flat people," he said, and they want to care for God's Green Earth. And some of it's about economics: "True champions are motivated for different reasons, but of course jobs and the economy are at the top of the list," he said.

The growth of Fleming's Conservatives for Clean Energy is testament to the growth in the number of Republicans and conservatives who are beginning to embrace the economic case for clean energy and climate action. Four years after he started the group, it has expanded to Virginia, South Carolina, Georgia, and Florida. Fleming admits that plans for future expansion to even more conservative states such as Alabama and Mississippi will be tough, but necessary, not just for economics or the future of the planet but for conservatives and Republicans to retain or regain political power.

"For us as conservatives to say there's not a problem and that the climate is not changing; that's just not a winning message," Fleming said. Instead, he added, the conservative message "has to be around economic growth. We have to say whenever you think about climate, there's this important economic sector—clean energy."

It's important to remember that Republicans once led environmental action and policymaking in America. Republican Teddy Roosevelt established five national parks, along with 150 national forests, fifty-one federal bird reserves, and eighteen national monuments, all covering more than two hundred million acres of public land. Richard Nixon signed into law fourteen bills that are now the bedrock of the nation's environmental legislation, including the National Environmental Policy Act, the Clean Air Act, and legislation that created the Environmental Protection Agency. George H. W. Bush ordered the establishment of the EPA's Energy Star program, authorized the first tax incentives for renewable energy, and signed into law the 1992 Energy Policy Act, which required states to set minimum commercial building efficiency codes.

Still, President Biden and his cabinet like to compare their plans to remake the American economy on clean energy and climate action to another bold and unprecedented initiative from another Democratic president. Sixty years before Biden outlined his climate and clean energy agenda in his first speech to a joint session of Congress, newly elected President John F. Kennedy stood in the same spot in the House rostrum to give his first speech to Congress. The country was recovering from a recession. The failed Bay of Pigs invasion just a month earlier—Kennedy's equivalent of Biden's messy withdrawal from Afghanistan—had stained the new president and sent his popularity ratings tumbling. And then on May 25, 1961, there was Kennedy standing before the Congress to ask it to approve the most expensive, most uncertain program the country had ever considered: A $531-million program to put a man on the moon within a decade.[18]

"Let me be clear . . . I am asking the Congress and the country to accept a firm commitment to a new course of action, a course which will last for many years and carry very heavy costs," Kennedy said. Despite the costs, "if we go only halfway, or reduce our sights in the face of difficulty, in my judgement it would be better not to go at all."

Kennedy's Apollo program that led to Neil Armstrong's first step on the lunar surface changed the arc of history for all of mankind, but it also profoundly changed the American economy. Solar panels, integrated circuits, digital telemetry, and, yes, freeze-dried food, all have roots in the Apollo program. Today, Americans rightly see Kennedy's moonshot program as one of the country's greatest and most impactful accomplishments and something that brought the country together regardless of politics. In a CBS News poll in 2019, more than sixty years after Kennedy famously told Congress and the world, "We choose to go to the moon," more than 80 percent of Americans surveyed said they still had a favorable opinion of the moonshot program. Nearly 75 percent said it was worth the cost and effort it took to get there.

That wasn't the case when Kennedy announced his plans. Back then, the majority of Americans were opposed to the program, seeing it as unnecessary and too costly. Congressional Republicans, still smarting from losing the White House to Kennedy and the Democrats, weren't happy about spending what was then a mind-bogglingly unprecedented amount. Plenty of scientists and other big thinkers of the day weren't convinced we could actually reach the moon, at least not in the mere ten years that Kennedy challenged the country to do so. And, of course, even decades after it happened, moon-landing deniers still float conspiracy theories that it never actually happened at all.

What sold the American people—including Kennedy's political opponents—on the moonshot program were two things. First and foremost was the threat from America's unequivocal foe of the time, communist Russia. If Russia had not put the world's first satellite into orbit with Sputnik in 1957 or put the first man into space when Soviet cosmonaut Yuri Gagarin orbited the Earth in April 1961, Kennedy would not have had the urgency or found the political or public support for the moonshot program. The second major factor was the economic case for the space race. Like Biden's climate agenda, Kennedy's moonshot program was also cloaked as an economic program to help get the country out of recession and bring down unemployment, which at the time of Kennedy's speech was about 7 percent. As an economic program, Apollo worked: At its peak, the program created more than four hundred thousand jobs across the country.[19] More than twenty thousand companies and contractors worked on the project. Businesses such as Boeing, McDonnell Douglas, Motorola, Westinghouse, Rockwell International, and IBM all have roots that run through the Apollo program. Those companies and their industries helped America leap ahead of the rest of the world into the age of technology in the years after Apollo, solidifying the country's standing as the most powerful economy in the world.

Federal and state policies that advance clean energy promise to similarly turn the Boeings and McDonnell Douglases of today—companies like Tesla, Rivian, First Solar, NextEra, GE Wind, ChargePoint and Proterra, to name a few—into tomorrow's behemoths, bringing up tens of thousands of smaller companies and benefiting hundreds of thousands of American workers with them.

"This is our generation's moonshot," Department of Energy Secretary Granholm said at Biden's international climate leaders' summit in April 2021. "Less than a decade after Kennedy declared our nation's choice to go to the moon, we planted an American flag on that cratered surface. Today, we choose to solve the climate crisis."[20]

Chapter Thirteen

The World and All of U.S.

More than one hundred world leaders had already given their opening speeches at the United Nations COP26 climate conference in Glasgow, Scotland, droning on one after another after another for a full day and a half at the beginning of November 2021. Leaders of the big countries—Biden, France's Emmanuel Macron, Germany's Angela Merkel—commanded the spotlight a day earlier. The press was beginning to turn its attention elsewhere.

And then, around lunchtime on the second day of the climate summit, the prime minister of the little country of Barbados, a woman in a white suit and red-framed glasses, took to the podium. In a poignant eight-minute, fifteen-second speech, she explained exactly what was really needed from the United States and other major countries to solve the global climate crisis.[1] "The pandemic has taught us that national solutions to global problems do not work," Prime Minister Mia Amor Mottley began. The same held true with climate change. But on that issue, she said, "there is a sword that can cut down this Gordian knot. It has been wielded before." That sword, she went on to explain, was money.

As Mottley reminded attendees and the rest of the world, the planet's wealthiest countries in 2009 promised to raise and distribute $100 billion annually by 2020 to smaller, developing countries like hers to expand clean energy and reduce fossil fuels ("climate mitigation," in policy circles) and take steps to protect their countries and their economies from climate change disasters ("climate adaptation"). Global warming and climate change, after all, are global problems that demand global solutions. Yet a dozen years later, the wealthy countries, including the United States, still had not lived up to their promises, as Mottley pointed out. By the time COP26 began, developed countries were still $20 billion shy of the $100 billion they had promised.

And now in Glasgow, they were saying they wouldn't reach the goal until 2023 at the earliest.

"Failure to provide the critical finance, and that of loss and damage is measured . . . in lives and livelihoods in our communities," Mottley said solemnly. "This is immoral. And it is unjust."

United Nations organizers had gone into the Glasgow climate summit with three overarching goals: To get wealthy countries to pony up the $100 billion in aid they had promised; to get all countries to agree to reduce their carbon emissions by at least 50 percent by 2030 in order to keep the Earth's warming below 1.5 degrees Celsius, and help communities and countries adapt to climate change.[2] As COP26 came to a close on November 13, 2021, none of those goals were met. Exhausted negotiators left only able to say they had made progress but acknowledged glumly they had not done enough to prevent a worsening of the climate crisis. To keep the Earth's temperature below the critical 1.5 degree Celsius mark, global greenhouse gas emissions need to decline by at least 45 percent by 2030. Instead, nothing negotiators had done in Glasgow changed the fact that global emissions remained on track to rise about 14 percent by 2030.

"Let me emphasize as strongly as I can: Job *not* done," U.S. special climate envoy John Kerry said at a press conference at the conclusion of the summit.[3]

The good news, however, is that some of the agreements and pledges made in Glasgow actually *could* get the job done. But they aren't the agreements and pledges made by countries or world leaders. The announcements and pledges that might matter most are those that came from businesses, banks, and financial institutions during COP26.

One of the lessons learned from the 2021 climate summit—and every COP before it—is that the world can't always depend on promises and pledges from countries, especially when they're voluntary. Any country's pledges and promises can be overturned by the shifting of political winds, like what happened in the United States during the Trump administration. They can fall short, as Barbados and the world learned from the $100 billion pledge from wealthy countries. And they can be forgotten in between meetings of bureaucrats that only happen once a year or so.

Of course, businesses break or come up short on their promises as well. But what drives businesses is something more powerful than any political promises or voluntary pledges: Money and economics. And when you augment the driving force of money and economics with the ability of governments, shareholders, customers, and employees to hold businesses accountable for their actions, that's what can make a difference on issues such as climate change.

From the start of the COP 26 summit,business and economics were a big part of the proceedings. That was unusual. At earlier U.N. climate summits, businesses were all but shunned and prohibited. They were viewed—and for valid reasons—as obstructionists, not catalysts, for climate action. That perception largely disappeared at Glasgow. Business leaders from bankers to clean-tech entrepreneurs were welcomed by country negotiators as partners, not adversaries. "We've never had as much corporate presence or commitment," said Kerry, who as a U.S. senator or diplomat has attended almost every U.N. climate summit.[4] One of the world's best-known business leaders, Bill Gates, noted that clean energy and clean-energy innovation "was literally on center stage" in Glasgow,[5] while clean-energy media entrepreneur and writer Joel Makower called COP26 "The Business COP."[6]

It was as if the world had finally recognized that climate change was now an economic issue.

While political negotiators stumbled in their quest to reach major new commitments to reduce carbon emissions at COP26, businesses and investors made one announcement after another to reduce carbon emissions by injecting private money into renewable energy, by withholding financing from coal and other fossil fuel operations, and by directing their sizeable investment funds to support only those companies helping the world reduce carbon emissions. If businesses and investors deliver on the commitments they made in Glasgow—and granted, that's a big if—they could make a bigger difference in catalyzing and speeding up the global economic shift to clean energy than any country pledge. Some examples:

- A group of investors led by the Rockefeller Foundation and the philanthropic arms of retailers IKEA and Amazon announced at COP26 that they would put up and pay up even if the governments of the wealthy nations did not. The new Global Energy Alliance for People and Planet announced by Rockefeller, the IKEA Foundation, and the Bezos Earth Fund pledged to raise and distribute $100 billion in capital to developing countries over the next decade to help them shift to clean energy.[7] According to the alliance, getting $100 billion to developing countries will help provide one billion people with clean, renewable energy, avert four billion tons of carbon emissions, and support 150 million jobs. "The world is undergoing an economic upheaval, in which the poorest are falling farther behind and being battered by climate change's effects," Rockefeller Foundation president Dr. Rajiv Shah said in announcing the alliance. "Green energy transitions with renewable electrification are the only way to restart economic progress for all while at the same time stopping the climate crisis."
- A coalition of 450 banks, insurers and investment and asset management firms across 45 countries announced a new alliance to leverage their combined $130 trillion in investment assets to fund projects designed to help

the world achieve net-zero emissions by 2050. The Glasgow Financial Alliance for Net Zero (GFANZ) is cochaired by former Bank of England governor and impact investor Mark Carney and Michael Bloomberg, the American businessman, politician, and philanthropist.[8] While critics of GFANZ pointed out that many of the banks and other investment firms still had major holdings in fossil fuel companies at the time of the announcement, GFANZ organizers said those would be phased out over time. "Winning the battle against climate change will require vast amounts of new investment," Bloomberg said in Glasgow. "And the majority will have to come from the private sector."

- Nearly seventy companies, including thirty-three of the world's largest financial institutions, agreed to quit funding coal projects as part of the Powering Past Coal Alliance, a public-private partnership group that seeks to phase out unabated coal power through changes in finance, contracts, and permitting.[9] "Climate change is the critical issue of our time, and it's clear that coal is incompatible with a net-zero future," Jenn-Hui Tan of financial giant Fidelity International said in joining the anticoal alliance. The alliance's goal to phase out coal was stronger than anything agreed to by the countries at COP26.
- A group of forty of the world's biggest cement and concrete makers, representing 80 percent of the total production outside of China, announced they would cut their carbon emissions by 25 percent by 2030 and eliminate carbon completely from their products by 2050. Cement and concrete producers are some of the biggest industrial emitters of carbon dioxide on the planet.[10] The announcement by the Global Cement and Concrete Association represented a huge shift in the industry's recognition of its role in battling climate change.

Still other business organizations are using their money to choke off fossil fuels and expand clean energy. More than thirty insurance companies announced they would quit underwriting insurance coverage for coal mines, coal-fired power plants, and railroad and trucking companies that transport coal. That means coal-related companies will have to pay much higher insurance premiums from the shrinking number of insurers who will work with them, driving up operating costs, or they face the possibility that they can't get financing for their projects because lenders will view them as too risky. "Without insurance there is no financing," Thomas Buberl, CEO of French insurance company AXA and the leader of the NetZero Insurance Alliance, told the *Washington Post.*[11]

Other public financing organizations also are backing away from dirty energy. At COP26, the Climate Investment Funds (CIF) organization that is affiliated with the World Bank launched new programs it said will invest $2.3 billion in developing countries to ease the transition away from coal and expand renewable energy.[12] The United States agreed to contribute $500 million per year in "green bonds" to fund those initiatives.

Announcing America's participation and contribution in the CIF programs was U.S. Treasury Secretary Janet Yellen. Her appearance in Glasgow was further affirmation that climate change is now an economic issue: It was the first time a U.S. treasury secretary ever attended a COP event.[13] "The reason I am here is because climate change is not just an environmental issue," Yellen said. "It is not just an energy issue. It is an economic, development and market-destabilizing issue. And I would not be doing my job if I did not treat it with the seriousness warranted."

Yellen also reminded attendees that the United States plans to quadruple its international climate finance for developing countries by 2024, to $11.4 billion per year, which could help slice a few strands of that "Gordian knot" that Barbados Prime Minister Mottley described. But what Yellen didn't mention was that U.S. commitment is still subject to congressional approval. And getting that might not be easy. Foreign climate finance is tricky. Paying other countries to convert to solar and wind and invest in seawalls and better ports and bridges will undoubtedly rub nationalists and conservatives in the United States the wrong way. If America can hardly pay for its own clean-energy transition, they say, why should we be responsible for paying for other countries to do so?

The answer lies in the fact that doing so will drive global economic growth, which is good for America and every country. It will open new opportunities for U.S. companies to invest overseas, which is good for keeping America competitive on the global stage. It will help blunt the economic costs of climate change that are rising each year in America and beyond. And it will help save the planet, which is sort of good for business too.

To understand how sending money to developing countries to mitigate and adapt to climate change would create economic benefits for the United States, think about soybeans.

Each year, the United States provides about $1.5 billion for foreign agricultural aid to developing countries, according to the U.S. Agency for International Development (USAID).[14] That money is used by developing countries to feed their people, either by investing in local farms or by purchasing agricultural goods produced in the United States. Annual U.S. agricultural exports to developing countries totals about $90 billion, according to USAID. That's about sixty times the amount of U.S. foreign agricultural aid given annually to those countries, a pretty good investment. It puts money in U.S. farmers' pockets and creates jobs all along the agricultural export supply chain, which according to USAID supports about 1.2 million U.S. jobs. Feeding the world also helps stabilize developing countries, reducing global conflict. As for soybeans? They're by far the biggest agriculture export from the United States, accounting for about $26 billion in exports every year.[15] When China or Mexico or Bangladesh or Barbados buys U.S. soybeans, they're helping soybean farmers in Iowa or Illinois or North Carolina and the entire U.S. agricultural industry.

Now instead of soybeans, think about electric vehicles or solar panels or batteries made in America. If we send money to other countries to help them buy more solar and wind or electric cars and bicycles—especially money that can be leveraged to attract additional private funding—it can help create jobs and drive growth at automakers like Ford and GM in Detroit or Kentucky. It can help solar panel manufacturers like First Solar in Ohio or LG Solar in Alabama. It can help energy efficient air conditioner makers in North Carolina and battery makers in Georgia. Likewise, when foreign countries shore up their seawalls and bridges to adapt to climate change, it can also mean business for U.S. engineering, construction, and construction supply companies who do business overseas. Reducing global greenhouse gas emissions also means reducing the economic costs of climate change for every country, since extreme heat in one part of the globe can mean intense hurricanes in another, just as severe droughts in one part of the world can impact crop prices in another.

In addition to economics, there's another reason businesses, investors, and industries are compelled to live up to their promises to invest more in clean energy and divest from dirty fuels: They have to answer to shareholders, customers, and increasingly, to regulators. Remember what Engine No. 1 did to shake up the board at Exxon Mobil, and what shareholder groups are increasingly doing in other countries across the economic spectrum. Remember what employee groups are doing to put pressure on their employers to stand up for climate action and their promises, as well as the "green halo" effect that makes consumers want to do business with companies that do the right thing on climate. And as Yellen reminded business leaders and other attendees at COP26, remember that—like other countries—the United States plans to require financial companies to increase climate-related disclosures and other climate-related data that's available to investors, lenders, regulators, and others.

Amid all the lofty ambitions and far-reaching goals at COP26, in the end, the final agreement hammered out between countries came down to the belabored parsing of a few important words. Near the end of the negotiations, representatives of nearly all the 190 countries at the climate summit were about to sign on to an agreement that said they would "phase out" unabated coal. But at the last minute, India and China balked. They would only agree to "phase *down*" coal. The small but significant change in wording incensed environmental groups and other countries, even though everybody involved knew that coal was on its way to being phased out not by government pledges but by economics.[16] In just about every part of the world, electricity from solar and wind is already cheaper than energy produced from coal, and renewable energy is only getting cheaper while coal is only getting more expensive. Economics is already doing what words on a paper cannot.

U.S special climate envoy John Kerry declared as much when he addressed the media at a press conference at the end of COP26. While the job was still not done, he said, progress had been made. He also specifically addressed the wording change around coal and provided his perspective on the choice that he and other negotiators had to make.

"We had a choice between whether or not we leave Glasgow with all these other things that we've accomplished and whether or not we change a word that still says . . . we've got to phase it down," Kerry said.[17] "I'll take 'phase it down' and fight next year and on through the next year as we get to where we need to go."

Yet as Kerry and other COP26 attendees were boarding their planes to leave Glasgow amid a job unfinished and more work remaining in the years to come, the urgency of climate change reared its head once again, and in truly terrifying ways. While negotiators in Scotland were parsing the words "phase out" or "phase down," a biblical combination of rainstorms, hail, flooding, and even snow was pummeling the Egyptian city of Aswan, one of the driest cities in the world that hardly ever gets rain. The wild weather killed at least three people. It also awoke and unleashed a mass invasion of thousands of burrowing scorpions that stung more than 450 people so badly they had to be hospitalized, according to Egyptian newspaper *Al-Ahram* and other news outlets.[18] Back in the United States, meanwhile, severe weather and heavy rains battered the Northeast, howling through New York City like a banshee and spawning a swarm of at least seven tornadoes that ripped up property and damaged homes and businesses stretching from New York to Connecticut and caused flight delays coming in from Glasgow and elsewhere. According to the National Weather Service, four tornadoes touched down on Long Island alone. It was the first time in history that tornadoes had ever been recorded in that area in the month of November.[19]

Chapter Fourteen

What We Don't (and Do) Know

On the island of Borneo, in the shadow of Mount Kinabalu, the third-highest island peak on Earth, jungle plants are digging their roots deep into the soil and doing the business of mining some of the most important elements necessary for our clean-energy future.

You read that right. The plants are doing the mining.[1]

It turns out that numerous species of fern-like plants and small trees, including *Alyssum murale*, *Phyllantus balgoyii*, and hundreds of others around the world, actually suck up nickel, cobalt, and other metals naturally as they grow, their tender roots acting like magnets for elements that otherwise could only be obtained by digging up and destroying the Earth, plants, and animals around them.[2] Scientists in Borneo, with the watchful eye of investors, entrepreneurs, and mining companies around the globe eagerly tracking their work, have now figured out how to cut back the growth shoots of such plants once or twice a year, burn or squeeze them, and extract nickel, a key component of electric car batteries and the generating equipment used in wind turbines.

Known as "phytomining," the process could someday produce enough nickel to supply the ever-increasing demand for battery energy storage while also abating the environmental destruction, the carbon emissions, and the millions of gallons of wasted water that come with traditional mining operations. It's an optimistic potential solution for an otherwise seemingly unsolvable problem—how to produce enough metals for all the batteries we need for energy storage and electric vehicles without destroying the planet while doing it. It also illustrates the importance of the interconnected web of government, scientists, business, and capitalism necessary to solve the climate crisis before it kills our economy. The process of phytomining was developed beginning in 1983, far from Borneo in the bowels of the obscure Washington, D.C., labs of the U.S. Department of Agriculture's Agricultural Research Service. Research agronomist Rufus Chaney came across the idea during the fifty-two years he spent at the lab, quietly studying such processes as part of his work researching how to clean up toxic Superfund sites.[3] Chaney retired in 2016. Now, venture capitalists, mining companies, and start-up entrepreneurs are looking for ways to commercialize Chaney's idea and the groundbreaking work—quite literally—being done in places like Borneo.

Perhaps most importantly, though, the idea of getting metals from plants to help make and store clean energy is illustrative of just how little we really know about the future of clean energy and climate solutions. The fact is, as recently as a decade or two ago, nobody in their right mind thought we could get enough energy from the sun or wind to power entire American cities, or that electric vehicles would take off so quickly that every major car manufacturer would blow up their business models, throw out their plans, and abandon vehicles powered by fossil fuels that drove the world's economy for more than a century. And the fact is, we just don't know what the future holds in terms of scientific and business breakthroughs that could alter the world's approach to climate change.

What we do know is we need to move quickly.

We also know groundwork being laid today will yield climate solutions that most of us can't even fathom today. That groundwork includes game-changing public policies coming out of Washington. It includes the innovation coming from climate-tech entrepreneurs and the venture capitalists funding them. And it includes the catalyzing public protest ringing across the country and throughout the world emphasizing the need to do more to save the planet.

What we also now know is that the rising economic costs of climate change are impacting everyone who lives on this planet. These costs are unsustainable. The United States might have the most powerful economy on the planet, but it can't continue to weather $145 billion annual economic damages from climate and weather-related disasters forever and continue to have the quality of life and the longevity of life that we have all come to expect. And if the United States can't afford the costs of climate change, what about smaller nations around the world—in other words, just about any other country? Beyond the borders of the United States, the economic costs of climate change and the economic benefits of climate action will reshape the global economy and determine the winners and losers during the next generation on Earth. Countries that embrace clean energy and climate action will benefit from more jobs, investments, and economic growth. Countries that ignore climate change or allow factions of their population and incumbent industries to keep them in the past will flounder.

We've also learned that most people, not just aging hippies or young people but businesspeople, farmers, and military leaders want climate action, and now. In the more than twenty years since my organization was founded by two environmentally minded Silicon Valley businesspeople, Nicole Lederer and Bob Epstein, the number of business leaders, investors, and other professionals who are a part of E2 has swelled from a handful of tech entrepreneurs to a national network of more than eleven thousand members and supporters from every walk of life. These include business leaders from every state and every sector of our economy who share the common realization that what's good for the environment is good for the economy, and vice versa.

In E2's work, and in my previous twenty-plus years as a business, technology, and political journalist, I've witnessed just how fast—and how slow—change can happen. My colleagues and I have seen how smart climate and environmental policy at the federal, state, and international levels can drive businesses and investments and create millions of jobs. We've seen how neither government nor business nor the general public alone can accomplish change. We've realized it takes the combination of all three, threaded together by the common fabric of crisis or opportunity, to really make change happen quickly.

There are a few other things we've learned over the years. Here are five of them:

1. **Policy matters, and policy works.**
   Policy works by helping create jobs, including the more than three million jobs in clean energy as of 2021. Many of these jobs can be traced to past policies from Washington, such as the 2009 American Recovery and Reinvestment Act, and policies from states, such as the renewable energy standards now in place in 30 states. Policy also works by creating and sending market signals to businesses and investors who every day look to Washington and beyond for guidance as to which way the country and its economy are headed. And policy works to make the world a better place. If it didn't, the smog would still be too thick in Los Angeles to go outside, the Cuyahoga River would still catch on fire, and we'd all still be driving cars that get no more than twelve miles per gallon burning leaded gasoline.
2. **Government can't do it alone. Neither can businesses.**
   Free-market proponents and conservatives often suggest that private industry can solve the world's problems. Leave it up to business, they say, and the natural rhythms of capitalism will automatically create the supply, the demand, and the change we need. But history shows us that doesn't work, at least not when it comes to monumental societal problems like climate change. In part, that's because the problems are too big, and in

part because there's nothing natural about American capitalism, which has been tweaked and tinkered with since the days of Alexander Hamilton and the country's founding. If businesses and capitalism alone could solve a problem like climate change, it would have done so by now. But as long as there is an uneven playing field between clean energy and fossil fuels, as long as there are outside forces such as deep-pocketed incumbent industries that can shape markets and politics, as long as decades-old subsidies for outdated industries continue to exist, business alone can't solve problems as big as climate change.

3. **We don't know all the answers yet. But we need to find them, and fast.** My time as a technology and business journalist eventually taught me to never second-guess the power of innovation. When Google's founders told me back in the early 2000s that someday they would electronically catalog every single thing on the Internet, I scoffed. When a cash-strapped Jeff Bezos told me and other journalists that someday consumers would flock to his then-struggling online store for more than just books, and that someday anybody would be able to buy just about anything on Amazon, I shook my head. And when Steve Jobs proclaimed to me and other reporters who gathered each year for Apple's product announcement events that someday the world would take pictures, listen to music, watch videos, and communicate in untold new ways with their pocket-sized cell phones, I didn't believe it. Years later, when I suggested to an oil and gas lobbyist in Washington, D.C., over lunch that cars soon would get more than fifty miles per gallon and that someday we'd all be driving electric vehicles, he laughed and gave me the same look I gave when listening to Jobs, Bezos, and Google's Sergey Brin. We don't know all the answers to the climate crisis yet, but we can find them in the government labs, in the tech company campuses, or maybe in the forests of places like Borneo. The real question is, will we find them soon enough?
4. **Most politicians want to do the right thing. They just need the right reason and the political will to do it.**
   In 2009, I left the business and technology beats to cover Congress and the White House for the *Atlanta Journal-Constitution* newspaper. Barack Obama had just been elected president, and Democrats also controlled both houses of Congress. At the time, most Republicans in Congress wanted nothing more than to see Democrats and Obama fail. Georgia Senator Johnny Isakson certainly fell into that camp. But Isakson, a moderate who was elected as a Republican in one of the more conservative states in the country despite the fact that he supported abortion rights and some limited gun control, was also realistic and pragmatic. A former real

estate agent and investor, he was one of the richest members of the Senate at the time, yet he still drove an aging Ford Escape hybrid, lived in the same modest Atlanta-area house he and wife Diane Davidson bought decades earlier, and favored off-the-rack suits. Isakson was one of my favorite politicians to be around, and I'm convinced he was a truly good guy who wanted to do the right thing on climate. "I'm the first person to tell you we should reduce our carbon footprint . . . and the truth of the matter is we all care about the environment," Isakson said on the Senate floor in March 2015.[4] "The environment and business should work in harmony with one another, rather than be adversaries and enemies." The problem is, Republican Minority Leader Senator Mitch McConnell made it the mission of congressional Republicans, Isakson included, to sink anything and everything President Obama and Democrats proposed, including the 2009 Waxman-Markey carbon cap-and-trade legislation that could have altered the arc of global warming and prevented some of the natural disasters in the years that followed. Isakson also faced plenty of pressure from his fossil fuel–powered hometown utility, Southern Company, to push back against climate regulations. Isakson retired from the Senate in 2019 as he struggled with Parkinson's disease, and died in December 2021.[5]

Both climate change and partisanship have worsened since Isakson's days. But so have the economic costs from hurricanes and other storms battering Georgia and the rest of the Eastern seaboard. At the same time, clean-energy jobs have grown dramatically in Georgia and every other state. And even the biggest corporate supporters of fossil fuels—with Atlanta-based Southern Co. near the top of the list—are now shifting to clean, renewable energy sources just about as fast as they can. In order to garner more widespread support for climate and clean-energy policies, we need to show companies and politicians how those policies are translating into savings, jobs, and a better future. We need to find ways to make it more comfortable for politicians to buck their leadership when their local economies and the fate of the planet are at stake. We need to give them a reason for doing the right thing—and listen to their opinions on how to do it. That sort of thinking is what led to the passage of President Biden's infrastructure bill in November 2021 with rare bipartisan support in both the House and Senate. Even though the bill was replete with Democrat-led climate-related spending for electric vehicles, grid modernization, and clean-energy innovation, thirteen Republican House members and nineteen Republican senators, including GOP Leader McConnell, voted for it. They did so because they knew it would mean jobs, investments, and opportunities in their states.

5. **Progress on climate change can be lost. Very quickly.**
   Nothing proves this more than the reign of Donald Trump. In four years, President Trump and his cabinet stacked with former oil, gas, and coal industry executives rolled back more than a hundred environmental, clean-energy, and clean-air regulations. He tried to expand oil and gas drilling off every coastline, in every state, and even in national parks. He derailed energy efficiency standards for everything from LED light bulbs to dishwashers. He revoked rules that prevented coal companies from dumping waste into rivers and streams and canceled limits on how much mercury and other toxins power plants could emit. And he pulled the country out of the Paris Agreement—albeit temporarily—making the United States the only country in the world to abandon the most important climate agreement in history. Fortunately for the planet, President Biden systematically began rolling back the Trump rollbacks the day he took office. But imagine how much better off the country—and the planet—could have been if Trump and Republicans in Congress had not delayed and dismantled climate action for four long years. Beginning with the November 2022 midterm elections, Americans will decide whether to elect politicians who will continue to address the economic costs of climate change and support policies that are expanding clean energy and the jobs that come with it, or elect politicians who want to take us backward and keep us shackled to fossil fuels and on the sidelines while the rest of the world benefits from the greatest economic transition in recent history.

During another election year, during another era, one of America's greatest presidents took to the podium on May 13, 1908, to speak to governors assembled from across the country. Theodore Roosevelt was a Republican in his last year in office. Six months later, the country would elect his hand-picked fellow Republican, William Taft, as his successor. Roosevelt had presided over one of the most economically transformational and successful periods in U.S. history. He also was a conservationist and outdoorsman who understood the economic value of the country's natural resources and their historic role in making America great but also the economic value of conserving those resources and the importance of planning for a different, better future.

"We have become great in a material sense because of the lavish use of our resources, and we have just reason to be proud of our growth," Roosevelt told the governors gathered before him.[6] "But the time has come to inquire seriously what will happen when our forests are gone, when the coal, the iron, the oil, and the gas are exhausted, when the soils shall have been still further impoverished and washed into the streams, polluting the rivers, denuding the fields, and obstructing navigation. These questions do not relate only to the next century or to the next generation."

Back then, climate change didn't threaten the world and every human and animal like it does now. And Roosevelt didn't have the hindsight we now have to understand the connection between burning coal, oil, and gas and the ruining of the planet. But Roosevelt understood the importance of not wasting America's resources. He understood decisions he and other national and world leaders made at the time would impact generations for centuries to come. And he understood that if they didn't make wise decisions, with both the economy and the environment in mind, it would ultimately spell disaster for both.

"One distinguishing characteristic of really civilized men is foresight," he said. "We have to, as a nation, exercise foresight for this nation in the future; and if we do not exercise that foresight, dark will be the future!"

A century or so later, the words of the Republican president known as the Wilderness Warrior[7] still ring true. Today, we have the hindsight to know what an economy built on fossil fuels and energy waste has done to the planet. We have the foresight to know climate change is killing our economy, and that we can save it—and our planet—by expanding clean energy and other climate solutions.

The question is, will we choose a future that's dark and unsustainable, or one that's bright and promising?

Notes

FOREWORD

1. https://www.iea.org/reports/world-energy-outlook-2021/executive-summary

CHAPTER 1

1. Clifford Krauss and Peter Eavis, "Climate Activists Defeat Exxon in Push for Clean Energy," *New York Times,* September 16, 2021, https://www.nytimes.com/2021/05/26/business/exxon-mobil-climate-change.html.

2. "2021 Proxy Statement," Chevron, May 26, 2021, 87, https://www.chevron.com/-/media/shared-media/documents/chevron-proxy-statement-2021.pdf.

3. Milieudefensie et al. v. Royal Dutch Shell judgement, May 26, 2021, https://uitspraken.rechtspraak.nl/inziendocument?id=ECLI:NL:RBDHA:2021:5339.

4. Neela Banerjee, John H. Cushman Jr., David Hasemyer, and Lisa Song, "Exxon: The Road Not Taken," *Inside Climate News*, 2016, https://insideclimatenews.org/project/exxon-the-road-not-taken/.

5. "Vanguard Investment Stewardship Insights," Vanguard, May 2021. https://about.vanguard.com/investment-stewardship/perspectives-and-commentary/Exxon_1663547_052021.pdf.

6. "Vanguard Investment Stewardship Insights," Vanguard, May 2021. https://about.vanguard.com/investment-stewardship/perspectives-and-commentary/Exxon_1663547_052021.pdf.

7. "Vote Bulletin: ExxonMobil Corporation," BlackRock, May 26, 2021, https://www.blackrock.com/corporate/literature/press-release/blk-vote-bulletin-exxon-may-2021.pdf.

8. Justin Baer and Dawn Lim, "The Hedge-Fund Manager Who Did Battle With Exxon—and Won," *Wall Street Journal*, June 12, 2021, https://www.wsj.com/articles/the-hedge-fund-manager-who-did-battle-with-exxonand-won-11623470420.

9. Baer and Lim, "The Hedge Fund Manager."

10. Schedule 14A Proxy Statement, Exxon Mobil Corp., May 2021, https://www.sec.gov/Archives/edgar/data/34088/000090266421001694/p21-0854prec14a.htm.

11. Steven Mufson, "Why has Andy Karsner frightened the mighty ExxonMobil?" *Washington Post,* June 19, 2021, https://www.washingtonpost.com/climate-environment/2021/06/19/exxon-board-karsner-engine1/.

12. "Billion-Dollar Weather and Climate Disasters," NOAA, https://www.ncdc.noaa.gov/billions/.

13. "2020 National Large Incident Year-to-Date Report," National Interagency Fire Center, December 21, 2020, https://web.archive.org/web/20201229021815/https://gacc.nifc.gov/sacc/predictive/intelligence/NationalLargeIncidentYTDReport.pdf.

14. *Oregonian* staff, "Oregon Wildfires Destroyed More Than 4,000 Homes. Here's Where," *Oregonian*, October 27, 2020, https://www.oregonlive.com/wildfires/2020/10/oregon-wildfires-destroyed-more-than-4000-homes-heres-where.html.

15. Stephanie Butzer, "How are Colorado's wildfires still burning in the snow?" November 26, 2020, *Thedenverchannel.com,* https://www.thedenverchannel.com/news/wildfire/how-are-colorados-wildfires-still-burning-in-the-snow.

16. Caroline Linton, "Lightning Siege hits California with nearly 12,000 strikes in a week," *CBS News*, August 22, 2020, https://www.cbsnews.com/news/lightning-siege-hits-california-with-nearly-12000-strikes-in-a-week-2020-08-22/.

17. Vincent Moleski, "First Ever Fire Tornado Warning Issued in CA," *The Sacramento Bee*, August 16, 2020, https://www.firehouse.com/operations-training/wildland/news/21150367/first-ever-fire-tornado-warning-issued-in-ca.

18. Colby Bermel, "Newsom: No patience for climate change deniers amid historic wildfires," *Politico*, September 8, 2020, https://www.politico.com/states/california/story/2020/09/08/newsom-no-patience-for-climate-deniers-amid-historic-heat-fires-1316014.

19. Resources for the Future, "Wildfires in the United States 101: Context and Consequences," July 30, 2021, https://www.rff.org/publications/explainers/wildfires-in-the-united-states-101-context-and-consequences/.

20. University of California, Irvine, "UCI, Tsinghua U.: California's 2018 wildfires caused $150 billion in damages," December 7, 2020, https://news.uci.edu/2020/12/07/uci-tsinghua-u-californias-2018-wildfires-caused-150-billion-in-damages/.

21. "2020 Atlantic Hurricane Season takes infamous top spot for busiest on record," NOAA, November 10, 2020, https://www.noaa.gov/news/2020-atlantic-hurricane-season-takes-infamous-top-spot-for-busiest-on-record.

22. Kevin Byrne, "Isaias created a tornado outbreak as it raced up the East Coast," Accuweather, August 7, 2020, https://www.accuweather.com/en/hurricane/isaias-created-a-tornado-outbreak-as-it-raced-up-the-east-coast/791095.

23. Ron Brackett and Jan Wesner Childs, "Tennessee Tornado Death Toll Rises; Several People Still Missing in Putnam County," Weather.com, March 4,

2020, https://weather.com/news/news/2020-03-03-tennessee-tornado-damage-deaths-severe-storms.

24. Jason Samenow, "Cedar Rapids and nearby Iowa communities, still in shambles days after destructive derecho, plead for help," *Washington Post*, August 14, 2020, https://www.washingtonpost.com/weather/2020/08/14/cedar-rapids-iowa-derecho/.

25. "2021 National Large Incident Year-to-Date Report," National Interagency Fire Center, December 12, 2021, https://gacc.nifc.gov/sacc/predictive/intelligence/NationalLargeIncidentYTDReport.pdf.

26. Matthew Cappuchi, "Denver still hasn't seen any snow this fall. That's a record," *Washington Post*, November 22, 2021, https://www.washingtonpost.com/weather/2021/11/22/denver-snow-record-dry/.

27. Alexandra Kelley, "Damage from Ida estimated to cost $18 billion," *The Hill,* September 1, 2021, https://thehill.com/changing-america/resilience/natural-disasters/570493-damage-from-ida-estimated-to-cost-18-billion.

28. *Inquirer* staff, "Ida death toll rises in Philly region," *Philadelphia Inquirer*, September 2, 2021, https://www.inquirer.com/weather/live/ida-philadelphia-flooding-tornado-pennsylvania-new-jersey-20210902.html.

29. NOAA, "Billion-dollar disasters," https://www.ncdc.noaa.gov/billions/.

30. "Fourth National Climate Assessment," NOAA et al., November 23, 2018, https://nca2018.globalchange.gov/.

31. Swiss RE, "The Economics of Climate Change," April 22, 2021, https://www.swissre.com/institute/research/topics-and-risk-dialogues/climate-and-natural-catastrophe-risk/expertise-publication-economics-of-climate-change.html.

32. World Meteorological Organization, United Nations Office for Disaster Risk Reduction, "Climate and weather-related disasters surge five-fold over 50 years," September 1, 2021, https://news.un.org/en/story/2021/09/1098662.

33. "Record hurricane season and major wildfires—the natural disaster figures for 2020," Munich RE, January 7, 2021, https://www.munichre.com/en/company/media-relations/media-information-and-corporate-news/media-information/2021/2020-natural-disasters-balance.html.

34. Michael Finney and Randall Yip, "California wildfires could lead to major spikes in cost of home insurance," *ABC 7 News,* October 7, 2020, https://abc7news.com/home-insurance-fire-fair-plan-risk-score/6797448/.

35. "Arctic sea ice has reached minimum extent for 2021," National Snow & Ice Data Center, September 22, 2021, https://nsidc.org/news/newsroom/arctic-sea-ice-has-reached-minimum-extent-2021.

36. Susanne Rust and Tony Barboza, "How climate change is fueling record-breaking California wildfires, heat and smog," *Los Angeles Times,* September 13, 2020, https://www.latimes.com/california/story/2020-09-13/climate-change-wildfires-california-west-coast.

37. Lazard, "Levelized Cost of Energy, Levelized Cost of Storage, and Levelized Cost of Hydrogen," October 28, 2021, https://www.lazard.com/perspective/levelized-cost-of-energy-levelized-cost-of-storage-and-levelized-cost-of-hydrogen/.

38. EPA, "Sources of Greenhouse Gas Emissions," https://www.epa.gov/ghgemissions/sources-greenhouse-gas-emissions.

39. Thomas Czigler et al., "Laying the foundation for zero-carbon cement," McKinsey & Co., May 14, 2020.

40. Global Cement and Concrete Association, "Concrete Future," October 2021, https://gccassociation.org/concretefuture/.

41. Matt Gough, "California's Cities Lead the Way to a Gas-Free Future," Sierra Club, July 22, 2021, https://www.sierraclub.org/articles/2021/07/californias-cities-lead-way-gas-free-future.

42. Kyle Welborn, "Q3 2021 AgTech Venture Capital Investment and Exit Round Up," *CropLife News,* October 13, 2021, https://www.croplife.com/precision/q3-2021-agtech-venture-capital-investment-and-exit-round-up/.

43. Yahoo Finance, 2020–2021, https://finance.yahoo.com/quote/fan/.

44. Svenja Telle, "COP26 and the climate finance bubble," *PitchBook,* October 30, 2021, https://pitchbook.com/news/articles/cop26-2021-climate-change-finance-bubble.

45. E2, "Clean Jobs America 2021," April 19, 2021, https://e2.org/reports/clean-jobs-america-2021/.

CHAPTER 2

1. Cpl. Shawn C. Rhodes, "Easy Company Carries on Tradition of Excellence," U.S. Marines, February 2, 2004, https://www.lejeune.marines.mil/News/Article/Article/511982/easy-company-carries-on-tradition-of-excellence/.

2. Jeff Schogol, "Marine recruits being evacuated from Parris Island ahead of Hurricane Matthew," *Marine Corps Times*, October 5, 2016, https://www.marinecorpstimes.com/news/your-marine-corps/2016/10/05/marine-recruits-being-evacuated-from-parris-island-ahead-of-hurricane-matthew/.

3. Shawn Snow, "Parris Island to evacuate ahead of Hurricane Florence," *Military Times*, September 10, 2018, https://www.militarytimes.com/news/your-marine-corps/2018/09/10/parris-island-to-evacuate-ahead-of-hurricane-florence/.

4. "Commanding General Major General Julian D. Alford," U.S. Marines, https://www.trngcmd.marines.mil/Leaders/Leaders-View/Article/1875079/major-general-julian-d-alford/.

5. Gina Harkins, "Lejeune Commander Fires Back at Critics After Declining to Evacuate Base," *Military.com*, September 12, 2018, https://www.military.com/daily-news/2018/09/12/lejeune-commander-fires-back-critics-after-declining-evacuate-base.html.

6. Ben Werner, "Camp Lejeune Marines May Shelter on Base During Hurricane Florence," *USNI News*, September 12, 2018, https://news.usni.org/2018/09/12/36499.

7. World Vision, "2018 Hurricane Florence: Facts, FAQs and how to help," 2018, https://www.worldvision.org/disaster-relief-news-stories/2018-hurricane-florence-facts.

8. Elsa Gillis, "Hurricane Florence aftermath: 1-year-old dies afer vehicle flooded by rising water*,"* *Atlanta Journal-Constitution*, September 17, 2018, https://www.ajc.com/weather/hurricane-florence-aftermath-year-old-missing-after-vehicle-flooded-rising-waters/Y9SAFhpnshVgmF5t9xSiaM/.

9. "Updated Estimates Show Florence Caused $17 Billion in Damage," North Carolina Governor Roy Cooper, October 31, 2018, https://governor.nc.gov/news/updated-estimates-show-florence-caused-17-billion-damage.

10. Shawn Snow, "$3.6 billion price tag to rebuild Lejeune buildings damaged by Hurricane Florence," *Marine Times*, December 12, 2018, https://www.marinecorpstimes.com/news/your-marine-corps/2018/12/12/36-billion-price-tag-to-rebuild-lejeune-buildings-damaged-by-hurricane-florence/.

11. Senate Armed Services Committee hearing, "Navy and Marine Corps Readiness," December 12, 2018, https://www.c-span.org/video/?455659-1/marine-corps-commandant-navy-secretary-testify-readiness.

12. Joseph Trevithick, "A Tornado Left the USAF With Only One Active E-4B 'Doomsday Plane' for months," *The Drive*, March 5, 2018, https://www.thedrive.com/the-war-zone/18996/a-tornado-left-the-usaf-with-only-one-active-e-4b-doomsday-plane-for-months.

13. Steve Liewer, "Flood Recovery at Offutt could cost $1 billion and take five years," *Omaha World Herald,* September 16, 2019, https://omaha.com/local/flood-recovery-at-offutt-could-cost-1-billion-and-take-five-years/article_8f4fff1a-ff4e-5265-bf73-3d6df6be94e8.html.

14. James Doubek, "Air Force Needs Almost $5 Billion to Recover Bases From Hurricane, Flood Damage," *NPR*, March 28, 2019, https://www.npr.org/2019/03/28/707506544/air-force-needs-almost-5-billion-to-recover-bases-from-hurricane-flood-damage.

15. Doubek, "Air Force Needs Almost $5 Billion."

16. Sen. Roger Wicker and Sen. Dan Sullivan, "Polar icebreakers are key to America's national interest," *Defense News*, October 19. 2020. https://www.defensenews.com/opinion/commentary/2020/10/19/polar-icebreakers-are-key-to-americas-national-interest/.

17. Colin P. Kelley et al., "Climate Change in the Fertile Crescent and implications of the recent Syrian drought," Proceedings of the National Academy of Sciences, March 2, 2015, https://www.pnas.org/content/112/11/3241.

18. Jim Garamone, "Military Must Be Ready For Climate Change, Hagel Says," *DOD News*, October 13, 2014, https://www.defense.gov/News/News-Stories/Article/Article/603441/military-must-be-ready-for-climate-change-hagel-says/.

19. Department of Defense, "2014 Climate Change Adaptation Roadmap," https://www.acq.osd.mil/eie/downloads/CCARprint_wForward_e.pdf.

20. Federal Register, "Promoting Energy Independence and Economic Growth," March 31, 2017, https://www.federalregister.gov/documents/2017/03/31/2017-06576/promoting-energy-independence-and-economic-growth.

21. Military.com, "Naval Station Norfolk Base Guide," https://www.military.com/base-guide/naval-station-norfolk.

22. Nicholas Kusnetz, "Rising Seas Are Flooding Norfolk Naval Base, and There's No Plan to Fix It," *Inside Climate News*, October 25, 2017, https://insideclimatenews.org/news/25102017/military-norfolk-naval-base-flooding-climate-change-sea-level-global-warming-virginia/.

23. Union of Concerned Scientists, "On the Front Lines of Rising Seas: Naval Station Norfolk, Virginia," July 15, 2016, https://www.ucsusa.org/resources/front-lines-rising-seas-naval-station-norfolk-virginia.

24. Peter Coutu, "Could flooding and sea level rise cost Hampton Roads a military base?" *Virginian-Pilot,* August 6, 2018, https://www.pilotonline.com/military/article_f2bc6da2-975a-11e8-8119-4368a84d4813.html.

25. Virginia Coastal Policy Center, "Economic Toll of Sea Level Rise Could Be Significant in Hampton Roads, Study Finds," November 16, 2016, https://law.wm.edu/news/stories/2016/economic-toll-of-sea-level-rise-could-be-significant-in-hampton-roads-study-finds.php.

26. Tamara Dietrich, "Study: Sea level rise has already cost Virginians $280 million in home values since 2005," *Daily Press,* September 10, 2018, https://www.dailypress.com/news/dp-nws-home-value-loss-20180910-story.html.

27. Brendan Ponton, "City of Hampton uses innovative tool to fight flooding," 3*WTKR*, December 3, 2020, https://www.wtkr.com/news/city-of-hampton-uses-innovative-tool-to-fight-flooding.

28. American Farm Bureau Federation, "Farm Bankruptcies Rise Again," October 30, 2019, https://www.fb.org/market-intel/farm-bankruptcies-rise-again.

29. Argus Research, "US farm debt declines on high crop prices, federal aid," July 9, 2021, https://www.argusmedia.com/en/news/2232923-us-farm-debt-declines-on-high-crop-prices-federal-aid.

30. Dan Charles, "Farmers Got a Government Bailout in 2020, Even Those Who Didn't Need It," *NPR*, December 30, 2020, https://www.npr.org/2020/12/30/949329557/farmers-got-a-government-bailout-in-2020-even-those-who-didnt-need-it.

31. USDA, Wildfire and Hurricane Indemnity Program-Plus (WHIP+), August 2020, https://www.fsa.usda.gov/Assets/USDA-FSA-Public/usdafiles/FactSheets/2019/wildfire-and-hurricane-indemnity-program-plus_whip_august_2020.pdf.

32. Andrew Crane-Droesch et al., "Climate Change Projected To Increase Cost of the Federal Crop Insurance Program," USDA, November 4, 2019, https://www.ers.usda.gov/amber-waves/2019/november/climate-change-projected-to-increase-cost-of-the-federal-crop-insurance-program-due-to-greater-insured-value-and-yield-variability/.

33. Matthew Schwartz, "Iowa Derecho This August Was Most Costly Thunderstorm Event in Modern U.S. History," *NPR*, October 18, 2020, https://www.npr.org/2020/10/18/925154035/iowa-derecho-this-august-was-most-costly-thunderstorm-event-in-modern-u-s-histor#:~:text=More%20Podcasts%20%26%20Shows-,This%20Summer's%20Iowa%20Thunderstorms%20Were%20Some%20Of%20The%20Costliest%20On,more%20expensive%20than%20some%20hurricanes.

34. USDA, "2020 State Agriculture Overview, North Carolina," https://www.nass.usda.gov/Quick_Stats/Ag_Overview/stateOverview.php?state=NORTH%20CAROLINA.

35. Andy Uhler, "Hurricane means fewer North Carolina sweet potatoes," *Marketplace*, November 22, 2018, https://www.marketplace.org/2018/11/22/nc-farmers-lick-their-wounds-after-hurricane-devastates-sweet-potato-crop/.

36. Mackensy Lunsford, "Florence will cost farmers more than $1 billion in lost crops and livestock," *Asheville Citizen-Times*, September 26, 2018, https://www.citizen-times.com/story/news/local/2018/09/26/agricultural-losses-become-clearer-eastern-north-carolina-recovers-hurricane-florence/1410031002/.

37. Philip A. Stayer, "Hurricane Florence and the poultry industry: Coping with the aftermath," *Poultry Health Today*, December 13, 2018, https://poultryhealthtoday.com/mobile/article/?id=6315.

38. Stayer, "Hurricane Florence and the poultry industry."

39. Erin Durkin, "North Carolina didn't like science on sea levels . . . so it passed a law against it," *The Guardian*, September 12, 2018, https://www.theguardian.com/us-news/2018/sep/12/north-carolina-didnt-like-science-on-sea-levels-so-passed-a-law-against-it.

40. WTVD, "NC town rejects solar farm, fearing it would suck up all the energy from the sun," *WTVD ABC 11*, December 14, 2015, https://abc11.com/sun-solar-panels-energy/1122081/.

41. C-SPAN, "House Natural Resources Committee Hearing on Climate Change," February 6, 2019, https://www.c-span.org/video/?457612-1/house-natural-resources-committee-hearing-climate-change.

42. *The Eagle*, "Weather Whys: The Dust Bowl," October 17, 2019, https://theeagle.com/townnews/agriculture/weather-whys-the-dust-bowl/article_abdd3290-8c11-11e2-a59a-0019bb2963f4.html.

43. Rick Silva, "The Humboldt Fire, 10 years later," *Enterprise-Record*, June 15, 2018, https://www.chicoer.com/2018/06/15/humboldt-fire-10-years-later/.

44. Maraya Cornell, "How catastrophic fires have ravaged through California," *National Geographic*, November 13, 2018, https://www.nationalgeographic.com/environment/article/how-california-fire-catastrophe-unfolded.

45. Doyle Rice, "USA had world's 3 costliest natural disasters in 2018, and Camp Fire was the worst," *USA Today*, January 8, 2019, https://www.usatoday.com/story/news/2019/01/08/natural-disasters-camp-fire-worlds-costliest-catastrophe-2018/2504865002/.

46. James Rufus Koren, "Insurer Merced went belly up after Camp fire. Here's what policyholders need to know," *Los Angeles Times*, December 4, 2018, https://www.latimes.com/business/la-fi-merced-insurance-customers-20181204-story.html.

47. Swiss RE Group: "Secondary perils to wreak evermore natural catastrophe devastation globally," April 10, 2019, https://www.swissre.com/media/news-releases/nr-20190410-sigma-2-2019.html.

48. Gretchen Frazee, "How climate change is changing your insurance," *PBS Newshour*, November 27, 2018, https://www.pbs.org/newshour/economy/making-sense/how-climate-change-is-changing-your-insurance.

49. Council of Insurance Agents and Brokers, "Q3 P/C Market Survey 2021," https://www.ciab.com/resources/q3-pc-market-survey-2021/.

50. Johnson Damian, "10 states with highest homeowners insurance rates*," Mortgage Professional America*, July 14, 2021, https://www.mpamag.com/us/mortgage-industry/guides/10-states-with-the-highest-homeowners-insurance-rates/260722.

51. Leslie Kaufman, "Millions in Fire-Ravaged California Risk Losing Home Insurance," *Bloomberg*, September 25, 2021, https://www.bloomberg.com/news/articles/2021-09-25/millions-in-fire-ravaged-california-risk-losing-home-insurance.

52. Kaufman, "Millions in Fire-Ravaged California."

53. Matt Stiles, "Firefighter overtime surged 65% in a decade, costing California $5 billion a year in wages," *Los Angeles Times*, December 8, 2019, https://www.latimes.com/california/story/2019-12-08/wildfire-firefighter-overtime-budget.

54. James F. Peltz, "PG&E to file for bankruptcy as wildfire costs hit $30 billion," *Los Angeles Times*, January 14, 2019, https://www.latimes.com/california/story/2019-12-08/wildfire-firefighter-overtime-budget.

55. Mario Koran, "California power outages could cost region more than $2bn, some experts say*," The Guardian*, October 11, 2019, https://www.theguardian.com/us-news/2019/oct/11/california-power-outages-cost-business-wildfires.

56. LeBron James (@kingjames), "Man, these LA fires aren't no joke," Twitter, October 28, 2019, https://twitter.com/kingjames/status/1188770703438278657.

57. Alex Wigglesworth and Joseph Serna, "California fire season shatters record with more than 4 million acres burned," *Los Angeles Times,* October 4, 2020, https://www.latimes.com/california/story/2020-10-04/california-fire-season-record-4-million-acres-burned.

58. NIFC.gov, https://www.nifc.gov/.

59. Michael Finney and Randall Yip, "California wildfires could lead to major spikes in cost of home insurance," *ABC7News,* October 7, 2020, https://abc7news.com/home-insurance-fire-fair-plan-risk-score/6797448/.

60. California Department of Insurance, "Commissioner Lara puts focus on solutions," October 20, 2020, https://www.insurance.ca.gov/0400-news/0100-press-releases/2020/release104-2020.cfm.

61. Amy O'Connor, "Florida Property Insurance Market Inches Closer to Crisis," *Insurance Journal*, October 29, 2020, https://www.insurancejournal.com/news/southeast/2020/10/29/588564.htm.

62. Office of Governor Gavin Newsom, "Ahead of Peak Fire Season, Governor Newsom Announces Surge in Firefighting Support," March 30, 2021, https://www.gov.ca.gov/2021/03/30/ahead-of-peak-fire-season-governor-newsom-announces-surge-in-firefighting-support-3-30-21/.

63. California Department of Insurance, October 20, 2020.

CHAPTER 3

1. Patrick Kingsley, "Trump Says California Can Learn from Finland on Fires. Is He Right?" *New York Times,* November 18, 2018, https://www.nytimes.com/2018/11/18/world/europe/finland-california-wildfires-trump-raking.html.

2. Jonathan Lemire et al., "Trump spurns science on climate: 'Don't think science knows,'" *Associated Press,* September 14, 2020, https://apnews.com/article/climate-climate-change-elections-joe-biden-campaigns-bd152cd786b58e45c61bebf2457f9930.

3. Jon Greenberg, "Donald Trump's ridiculous link between cancer, wind turbines," *Politifact*, April 8, 2019, https://www.politifact.com/factchecks/2019/apr/08/donald-trump/republicans-dismiss-trumps-windmill-and-cancer-cla/.

4. IRENA, "Renewables increasingly beat even cheapest coal competitors on cost," June 2, 2020, https://www.irena.org/newsroom/pressreleases/2020/Jun/Renewables-Increasingly-Beat-Even-Cheapest-Coal-Competitors-on-Cost.

5. Eric Lipton, "'The Coal Industry Is Back,' Trump Proclaimed. It Wasn't," *New York Times,* October 5, 2020, https://www.nytimes.com/2020/10/05/us/politics/trump-coal-industry.html.

6. E2, "Clean Jobs America 2020," https://e2.org/reports/clean-jobs-america-2020/.

7. U.S. Bureau of Labor Statistics, "Solar and wind generation occupations: a look at the next decade," February 2021, https://www.bls.gov/opub/btn/volume-10/solar-and-wind-generation-occupations-a-look-at-the-next-decade.htm.

8. Center for American Progress, "Trump Administration Is Selling Western Wildlife Corridors to Oil and Gas industry," February 14, 2019, https://americanprogress.org/article/trump-administration-selling-western-wildlife-corridors-oil-gas-industry/.

9. Wes Siler, "The David Bernhardt Scandal Tracker," *Outside*, April 19, 2019, https://www.outsideonline.com/outdoor-adventure/environment/david-bernhardt-scandal-tracker/.

10. Steven Mufson, "Trump's pick for managing federal lands doesn't believe the government should have any," *Washington Post,* July 31, 2019, https://www.washingtonpost.com/climate-environment/trumps-pick-for-managing-federal-lands-doesnt-believe-the-government-should-have-any/2019/07/31/0bc1118c-b2cf-11e9-8949-5f36ff92706e_story.html.

11. Jack Fitzpatrick, "Interior's Pick to Lead Offshore Oil Regulator Has Industry Ties," *Morning Consult,* May 22, 2017, https://morningconsult.com/2017/05/22/interiors-pick-lead-offshore-oil-regulator-industry-ties/.

12. Texas Public Policy Foundation, "Doug Domenech joins TPPF as Director of Fueling Freedom Project," March 30, 2015, https://www.texaspolicy.com/press/doug-domenech-joins-tppf-as-director-of-fueling-freedom-project.

13. Jeff Turrentine, "Conflicts of Interest Swamp the Department of Interior," *NRDC onEarth*, December 1, 2017, https://www.nrdc.org/onearth/conflicts-interest-swamp-department-interior.

14. S&P Global, "Investment in US clean energy to total $55 billion in 2020," November 23, 2020, https://www.spglobal.com/platts/en/market-insights/latest-news/electric-power/112320-investment-in-us-clean-energy-to-total-55-bil-in-2020-generate-capital.

15. Wells Fargo, "Wells Fargo surpasses $10 billion in tax-equity financing of renewable energy projects," February 25, 2021, https://stories.wf.com/wells-fargo-surpasses-10-billion-in-tax-equity-financing-of-renewable-energy-projects/.

16. NextEra Energy, "Did You Know?" 2021, https://www.nexteraenergy.com/company/subsidiaries.html.

17. Associated Press, "Radioactive materials from nuclear plant leaking into Biscayne Bay, study says," *Orlando Sentinel*, March 8, 2016, https://www.orlandosentinel.com/features/gone-viral/os-turkey-point-nuclear-radioactive-biscayne-bay-story.html.

18. David Roberts, "Florida's outrageously deceptive solar ballot initiative, explained," *Vox*, November 8, 2016, https://www.vox.com/science-and-health/2016/11/4/13485164/florida-amendment-1-explained.

19. NextEra, https://www.nexteraenergy.com/.

20. Shariq Khan and Scott DiSavino, "NextEra doubles down on renewables as energy transition gathers steam," *Reuters*, January 26, 2021, https://www.reuters.com/article/us-nextera-energy-results/nextera-doubles-down-on-renewables-as-energy-transition-gathers-steam-idUSKBN29V1LM.

21. Companiesmarketcap.com, "Largest American companies by market capitalization," November 27, 2021, https://companiesmarketcap.com/usa/largest-companies-in-the-usa-by-market-cap/.

22. Abigail Johnson Hess, "How Tesla and Elon Musk became household names," CNBC, November 21, 2017, https://www.cnbc.com/2017/11/21/how-tesla-and-elon-musk-became-household-names.html.

23. Tesla, "Tesla Repays Department of Energy Loan Nine Years Early," May 22, 2013, https://www.tesla.com/blog/tesla-repays-department-energy-loan-nine-years-early.

24. U.S. Energy Information Administration, "Renewables account for most new U.S. electricity generating capacity in 2021," January 11, 2021, https://www.eia.gov/todayinenergy/detail.php?id=46416.

25. E2, "Clean Jobs America 2020," April 15, 2021, https://e2.org/reports/clean-jobs-america-2020/.

26. E2, "Clean Jobs America 2021," April 19, 2021, https://e2.org/reports/clean-jobs-america-2021/.

27. Travis M. Andrews, "Kentucky Coal Mining Museum in Harlan County switches to solar power," *Washington Post*, April 6, 2017, https://www.washingtonpost.com/news/morning-mix/wp/2017/04/06/the-coal-mining-museum-in-harlan-county-ky-switches-to-solar-power/.

28. Erica Joiner West, "Hitachi Automotive to make electric motors in Kentucky," Southern Automotive Alliance, January 21, 2021, https://southernautomotivealliance.com/hitachi-automotive-to-make-electric-motors-in-kentucky/.

29. Ford Motor Co., "Ford to Lead America's Shift to Electric Vehicles with New Mega Campus," September 27, 2021, https://media.ford.com/content/fordmedia/fna/us/en/news/2021/09/27/ford-to-lead-americas-shift-to-electric-vehicles.html.

30. E2, "Clean Energy Jobs by Congressional District," https://cleanjobsamerica.e2.org/.

31. Associated Press, "7,000 working at Tesla Gigafactory east of Reno," *Las Vegas Sun,* December 5, 2018, https://lasvegassun.com/news/2018/dec/05/7000-working-at-tesla-gigafactory-east-of-reno/.

32. Capital Dynamics, Switch, "Switch and Capital Dynamics Break Ground on Massive Solar and Battery Storage Developments," July 22, 2020, https://www.prnewswire.com/news-releases/switch-and-capital-dynamics-break-ground-on-massive-solar-and-battery-storage-developments-advancing-rob-roys-gigawatt-nevada-301097874.html.

33. E2, "Clean Jobs America 2021."

34. Energy Optimizers, "Monroe Local Schools," https://energyoptusa.com/fccase studies/monroe-local-schools/.

35. U.S. Department of Energy, "Energy Star Impacts," https://www.energystar.gov/about/origins_mission/impacts.

36. Ariel Fan interview, November 25, 2020.

37. E2, "Help Wanted: Diversity in Clean Energy," September 9, 2021, https://e2.org/reports/diversity-in-clean-energy-2021/.

CHAPTER 4

1. Chris Mooney, "30 years ago, scientists warned Congress on global warming," *Washington Post*, June 11, 2016, https://www.washingtonpost.com/news/energy-environment/wp/2016/06/11/30-years-ago-scientists-warned-congress-on-global-warming-what-they-said-sounds-eerily-familiar/.

2. EPA, "Can We Delay A Greenhouse Warming?" September 1983, https://nepis.epa.gov/Exe/ZyNET.exe/9101HEAX.TXT?ZyActionD=ZyDocument&Client=EPA&Index=1981+Thru+1985&Docs=&Query=&Time=&EndTime=&SearchMethod=1&TocRestrict=n&Toc=&TocEntry=&QField=&QFieldYear=&QFieldMonth=&QFieldDay=&IntQFieldOp=0&ExtQFieldOp=0&XmlQuery=&File=D%3A%5Czyfiles%5CIndex%20Data%5C81thru85%5CTxt%5C00000024%5C9101HEAX.txt&User=ANONYMOUS&Password=anonymous&SortMethod=h%7C-&MaximumDocuments=1&FuzzyDegree=0&ImageQuality=r75g8/r75g8/x150y150g16/i425&Display=hpfr&DefSeekPage=x&SearchBack=ZyActionL&Back=ZyActionS&BackDesc=Results%20page&MaximumPages=1&ZyEntry=1&SeekPage=x&ZyPURL.

3. Walter Sullivan, "Warming Trend Seen in Climate," *New York Times*, August 14, 1975, https://www.nytimes.com/1975/08/14/archives/warming-trend-seen-in-climate-two-articles-counter-view-that-cold.html.

4. Walter Sullivan, "Low Ozone Level Found Above Antarctica," *New York Times*, November 7, 1985, https://www.nytimes.com/1985/11/07/us/low-ozone-level-found-above-antarctica.html.

5. Carnegie Mellon University, "Hearing on Ozone Depletion and the Greenhouse Effect, Opening Statement," http://digitalcollections.library.cmu.edu/awweb/awarchive?type=file&item=437988.

6. Congress.gov, "S.2891 Global Climate Protection Act of 1986," September 29, 1986, https://www.congress.gov/bill/99th-congress/senate-bill/2891.

7. Congressional Record, Vol. 132, Part 19, September 29, 1986, https://www.congress.gov/bound-congressional-record/1986/09/29/senate-section?q=%7B%22search%22%3A%22global+warming+protection%22%7D&s=4&r=3.

8. Congressional Record, Vol. 133, Part 2, January 29, 1987, https://www.congress.gov/bound-congressional-record/1987/01/29/senate-section.

9. Robert Gillette, "Suggests Wearing Hats, Sunscreen Instead of Saving Ozone Layer: Hodel Proposal Irks Environmentalists," *Los Angeles Times*, May 30, 1987, https://www.latimes.com/archives/la-xpm-1987-05-30-mn-3572-story.html.

10. Dominick Mastrangelo, "Ron Johnson: Climate change is 'bulls---,'" *The Hill*, July 7, 2021, https://thehill.com/homenews/senate/561805-ron-johnson-climate-change-is-bullsh.

11. Jason Plautz and *National Journal*, "How Green Is Joe Biden?" *The Atlantic*, September 16, 2015, https://www.theatlantic.com/politics/archive/2015/09/how-green-is-joe-biden/452970/.

12. Lydia Saad, "Economy Reigns Supreme for Voters," Gallup, October 29, 2008, https://news.gallup.com/poll/111586/economy-reigns-supreme-voters.aspx.

13. Zach Hrynowski, "Several Issues Tie as Most Important in 2020 Election," Gallup, January 13, 2020, https://news.gallup.com/poll/276932/several-issues-tie-important-2020-election.aspx.

14. Alec Tyson and Brian Kennedy, "Two-Thirds of Americans Think Government Should Do More on Climate," Pew Research, June 23, 2020, https://www.pewresearch.org/science/2020/06/23/two-thirds-of-americans-think-government-should-do-more-on-climate/.

15. Alec Tyson, "How important is climate change to voters in the 2020 election?" Pew Research, October 6, 2020, https://www.pewresearch.org/fact-tank/2020/10/06/how-important-is-climate-change-to-voters-in-the-2020-election/.

16. Anthony Salvanto, "CBS News poll: Eye on Earth," April 19, 2021, https://www.cbsnews.com/news/climate-change-economy-opinion-poll/.

17. Lauren Camera, "Young Voters Not Excited About Joe Biden," *US News*, September 16, 2019, https://www.usnews.com/news/elections/articles/2019-09-16/young-voters-not-excited-about-joe-biden.

18. Julia Conley, "Sanders Scores Highest Mark on Sunrise Movement's Climate Report Card While Biden Told It's 'Parent-Teacher Conference Time,'" Common Dreams, December 5, 2019, https://www.commondreams.org/news/2019/12/05/sanders-scores-highest-mark-sunrise-movements-climate-report-card-while-biden-told.

19. Stefanie Feldman, "Surviving On Their Own: How Leaders in Appalachia's Remote Rural Communities Make Economic Development Policy," Duke University, December 4, 2009, https://dukespace.lib.duke.edu/dspace/handle/10161/1690.

20. C-SPAN, "President-elect Biden's Policy Agenda," December 8, 2020, https://www.c-span.org/video/?507096-1/president-elect-bidens-policy-agenda.

21. C-SPAN, "Joe Biden Remarks on Energy Policy in Wilmington, Delaware," July 14, 2020, https://www.c-span.org/video/?473845-1/joe-biden-remarks-energy-policy-wilmington-delaware.

CHAPTER 5

1. Larry Fink, "A Fundamental Reshaping of Finance," BlackRock, January 2020, https://www.blackrock.com/corporate/investor-relations/2020-larry-fink-ceo-letter.

2. Delta, "Delta commits $1 billion to become first carbon neutral airline globally," February 14, 2020, https://news.delta.com/delta-commits-1-billion-become-first-carbon-neutral-airline-globally.

3. Marc Benioff and Jane Goodall, "The Trillion Tree challenge that could save the planet," World Economic Forum, September 22, 2020, https://www.weforum.org/agenda/2020/09/planting-1-trillion-trees-save-the-planet-sustainability/.

4. Brad Smith, "Microsoft will be carbon negative by 2030," Official Microsoft Blog, January 16, 2020, https://blogs.microsoft.com/blog/2020/01/16/microsoft-will-be-carbon-negative-by-2030/.

5. David Gelles, "Microsoft Leads Movement to Offset Emissions With Internal Carbon Tax," *New York Times,* September 26, 2015, https://www.nytimes.com/2015/09/27/business/energy-environment/microsoft-leads-movement-to-offset-emissions-with-internal-carbon-tax.html.

6. Jacob Little, "Microsoft's Carbon Fee Model Presented by Alumna Tamara DiCaprio," *Daily Nexus,* October 16, 2015, https://dailynexus.com/2015-10-16/microsofts-carbon-fee-model-presented-by-alumna-tamara-dicaprio/.

7. Amazon, "Amazon is making big global investments in renewable energy," April 19, 2021, https://www.aboutamazon.com/news/sustainability/amazon-is-making-big-global-investments-in-renewable-energy.

8. Michael Wayland, "General Motors plans to exclusively offer electric vehicles by 2035," *CNBC*, January 28, 2021, https://www.cnbc.com/2021/01/28/general-motors-plans-to-exclusively-offer-electric-vehicles-by-2035.html.

9. Science Based Targets, https://sciencebasedtargets.org/.

10. Jane Fraser, "Citi's Commitment to Net Zero by 2050," Citigroup, March 1, 2021, https://blog.citigroup.com/2021/03/citis-commitment-to-net-zero-by-2050/.

11. David Benoit, "Citigroup's Jane Fraser Expects to Shed Some Clients for Climate Purposes," *Wall Street Journal,* December 7, 2021, https://www.wsj.com/articles/citigroups-jane-fraser-expects-to-shed-some-clients-for-climate-purposes-11638912251.

12. Alyssa Stankiewicz, "The Number of New Sustainable Funds Hits an All-Time Record," *Morningstar*, October 28, 2021, https://www.morningstar.com/articles/1062299/the-number-of-new-sustainable-funds-hits-an-all-time-record.

13. Tim Quinson, "Wall Street's $22 Trillion Carbon Time Bomb," Bloomberg News, *Financial Post*, November 24, 2021, https://financialpost.com/pmn/business-pmn/wall-streets-22-trillion-carbon-time-bomb.

14. Moody's Investors Service, "Research Announcement: Moody's—Financial firms that take rapid, predictable pace to zero financed emissions will win the race," October 12, 2021, https://www.moodys.com/research/Moodys-Financial-firms-that-take-rapid-predictable-pace-to-zero--PBC_1305598.

15. CERES, "Investors seek greater climate action in 2021 Proxy Season," April 2, 2021, https://www.ceres.org/news-center/press-releases/investors-seek-greater-climate-action-2021-proxy-season.

16. Matt Wirz, "British Hedge Fund Billionaire Takes Climate Fight to S&P 500," *Wall Street Journal,* January 28, 2021, https://www.wsj.com/articles/british-hedge-fund-billionaire-takes-climate-fight-to-s-p-500-11611842401.

17. As You Sow, "92% of Sysco Shareholders Support Net-Zero Climate Targets," November 29, 2021, https://www.asyousow.org/press-releases/2021/11/29/sysco-shareholders-support-net-zero-climate-targets.

18. Dwight Adams, "RFRA: Why the 'religious freedom law' signed by Mike Pence was so controversial," *IndyStar*, April 25, 2018, https://www.indystar.com/story/news/2018/04/25/rfra-indiana-why-law-signed-mike-pence-so-controversial/546411002/.

19. Associated Press, 'Bathroom bill' to cost North Carolina $3.76 billion," *CNBC*, March 27, 2017, https://www.cnbc.com/2017/03/27/bathroom-bill-to-cost-north-carolina-376-billion.html.

20. ClimateVoice, "It's Go Time," https://climatevoice.org/campaigns/.

21. E2, "Nearly 400 Business Leaders to Biden: Cut Emissions 50% by 2030," April 12, 2021, https://e2.org/releases/nearly-400-business-leaders-to-biden-cut-emissions-50-by-2030/.

22. CERES, "411 Businesses and Investors Support U.S. Federal Climate Target in Open Letter to President Biden," April 13, 2021, https://www.ceres.org/news-center/press-releases/411-businesses-and-investors-support-us-federal-climate-target-open.

23. World Economic Forum, "The Global Risks Report 2021," January 19, 2021, https://www.weforum.org/reports/the-global-risks-report-2021.

24. Nadja Popovich et al., "The Trump Administration Rolled Back More Than 100 Environmental Rules. Here's the Full List," *New York Times*, January 20, 2021, https://www.nytimes.com/interactive/2020/climate/trump-environment-rollbacks-list.html.

CHAPTER 6

1. C-SPAN, "White House Jobs Summit Opening," December 3, 2009, https://www.c-span.org/video/?290419-1/white-house-jobs-summit-opening.

2. NOAA, "2020 Atlantic Hurricane Season takes infamous top spot for busiest on record," November 10, 2020, https://www.noaa.gov/news/2020-atlantic-hurricane-season-takes-infamous-top-spot-for-busiest-on-record.

3. World Resources Institute, "America's New Climate Economy: A Comprehensive Guide to the Economic Benefits of Climate Policy in the United States," July 28, 2020, https://www.wri.org/research/americas-new-climate-economy-comprehensive-guide-economic-benefits-climate-policy-united.

4. Stanford University, "COVID lockdown causes record drop in carbon emissions for 2020," *Stanford Earth Matters*, December 10, 2020, https://earth.stanford.edu/news/covid-lockdown-causes-record-drop-carbon-emissions-2020#gs.h84938.

5. E2, "Clean Jobs, Better Jobs," https://e2.org/reports/clean-jobs-better-jobs/.

6. Ella Nilsen, "Joe Biden and Bernie Sanders are building new, policy-focused task forces," *Vox*, May 13, 2020, https://www.vox.com/2020/5/13/21257078/joe-biden-bernie-sanders-joint-unity-task-forces-democratic-policy.

7. SEIA.org, "North Carolina Solar," https://www.seia.org/states-map.

8. U.S. Department of Energy, "Land-Based Wind Market Report: 2021 Edition," https://www.energy.gov/sites/default/files/2021-08/Land-Based%20Wind%20Market%20Report%202021%20Edition_Full%20Report_FINAL.pdf.

CHAPTER 7

1. Justin Fishel, "John Kerry Signs Climate Deal With Granddaughter Seated In Lap," *ABC News,* April 22, 2016, https://abcnews.go.com/International/john-kerry-signs-climate-deal-granddaughter-seated-lap/story?id=38600097.

2. Congress.gov, "S.1733 – Clean Energy Jobs and American Power Act," February 2, 2010, https://www.congress.gov/bill/111th-congress/senate-bill/1733.

3. Jay Turner, "New EPA head Gina McCarthy proud of local roots," *Canton Citizen*, May 8, 2014, https://www.thecantoncitizen.com/2014/05/08/gina-mccarthy-profile/.

4. White House, "Readout of the First National Climate Task Force Meeting," February 11, 2021, https://www.whitehouse.gov/briefing-room/statements-releases/2021/02/11/readout-of-the-first-national-climate-task-force-meeting/.

5. Granholm, Jennifer (@JenGranholm), "My sincere thanks," Twitter, February 25, 2021, https://twitter.com/jengranholm/status/1365001115893309441?lang=en.

6. YouTube, "Gina McCarthy and Secretary Granholm Hit the Road in an EV," July 1, 2021, https://www.youtube.com/watch?v=g9pElxQpiQ8.

7. Sarah Holder, "Pete Buttigieg's Climate Vision: Local Fixes for a Planet in Crisis," *Bloomberg*, November 4, 2019, https://www.bloomberg.com/news/articles/2019-11-04/inside-pete-buttigieg-s-2-trillion-climate-plan.

8. Katie Surma, "Mary Nichols Was the Early Favorite to Run Biden's EPA, Before She Became a 'Casualty,'" *Inside Climate News*, February 9, 2021, https://insideclimatenews.org/news/09022021/mary-nichols-epa-joe-biden-michael-regan-environmental-justice/.

9. Catherine Morehouse, "Three's company: New Mexico joins California, Hawaii in approving 100% clean energy mandate," *Utility Dive*, March 13, 2019, https://www.utilitydive.com/news/threes-company-new-mexico-joins-california-hawaii-in-approving-100-clea/550390/.

10. *Forbes*, https://www.forbes.com/profile/janet-yellen/?sh=4f96799b36d7.

11. U.S. Department of the Treasury, "Remarks by Secretary of the Treasury Janet L. Yellen at the Venice International Conference on Climate," July 11, 2021, https://home.treasury.gov/news/press-releases/jy0271.

12. Clinton White House Archives, "Testimony of Dr. Janet Yellen, Chair, Council of Economic Advisers," March 4, 1998, https://clintonwhitehouse2.archives.gov/WH/EOP/CEA/html/19980304.html.

13. U.S. Department of the Treasury, "Treasury Announcs Coordinated Climate Policy Strategy with New Treasury Climate Hub and Climate Counselor," April 19, 2021, https://home.treasury.gov/news/press-releases/jy0134.

14. U.S. Department of the Treasury, "Readout of Financial Stability Oversight Council Meeting on July 16, 2021," https://home.treasury.gov/news/press-releases/jy0278.

15. U.S. Department of the Treasury, "Remarks by Secretary of the Treasury Janet L. Yellen on the Executive Order on Climate-Related Financial Risks," May 20, 2021, https://home.treasury.gov/news/press-releases/jy0190.

16. Marianne Lavelle, "Rick Perry Denies Climate Change Role of CO2," *Inside Climate News*, June 19, 2017, https://insideclimatenews.org/news/19062017/rick-perry-dismisses-co2-climate-change-oceans-environment-cnbc-kernen/.

17. Steve Mufson, "Scott Pruitt's likely successor has long lobbying history on issues before the EPA," *Washington Post*, July 5, 2018, https://www.washingtonpost.com/business/economy/epas-acting-administrator-has-long-lobbying-record-on-issues-before-the-agency/2018/07/05/a591cd40-6a6b-11e8-bea7-c8eb28bc52b1_story.html.

18. Heritage Foundation, "Elaine Chao, Former Distinguished Fellow," https://www.heritage.org/staff/elaine-chao.

CHAPTER 8

1. Obama White House, "Remarks by the President in State of the Union Address," January 25, 2011, https://obamawhitehouse.archives.gov/the-press-office/2011/01/25/remarks-president-state-union-address.

2. The White House, "Remarks by President Biden in Address to a Joint Session of Congress," April 29, 2021, https://www.whitehouse.gov/briefing-room/speeches-remarks/2021/04/29/remarks-by-president-biden-in-address-to-a-joint-session-of-congress/.

3. Erum Salam, "What's actually in Biden's Build Back Better bill? And how would it affect you?" *The Guardian*, October 18, 2021, https://www.theguardian.com/us-news/2021/oct/18/what-is-build-back-better-crash-course.

4. Congressional Research Service, "The Clean Electricity Performance Program (CEPP): In Brief," October 7, 2021, https://sgp.fas.org/crs/misc/R46934.pdf.

5. Coalition for Green Capital, "Clean Energy Accelerator," https://coalitionforgreencapital.com/accelerator/.

6. Congressional Research Service, "The Renewable Electricity Production Tax Credit: In Brief," April 29, 2020, https://sgp.fas.org/crs/misc/R43453.pdf.

7. Congressional Research Service, "The Energy Credit or Energy Investment Tax Credit (ITC)," April 23, 2021, https://crsreports.congress.gov/product/pdf/IF/IF10479.

8. U.S. Department of Energy, "DOE Announced Up to $8.25 Billion in Loans to Enhance Electrical Transmission Nationwide," April 27, 2021, https://www.energy

.gov/articles/doe-announces-825-billion-loans-enhance-electrical-transmission-nationwide.

9. U.S. Department of Transportation, "US Department of Transportation announces $180 million funding opportunity for low or no emission transit vehicles & facilities," February 11, 2021, https://www.transit.dot.gov/about/news/us-department-transportation-announces-180-million-funding-opportunity-low-or-no.

10. U.S. Environmental Protection Agency, "EPA Awards $10.5 Million to Clean Up 473 School Buses in 40 States," April 19, 2021, https://www.epa.gov/newsreleases/epa-awards-105-million-clean-473-school-buses-40-states.

11. U.S. Department of the Treasury, "Treasury Announces Coordinated Climate Policy Strategy With New Treasury Climate Hub and Climate Counselor," April 19, 2021, https://home.treasury.gov/news/press-releases/jy0134.

12. The White House, "Leaders Summit on Climate Summary of Proceedings," April 23, 2021, https://www.whitehouse.gov/briefing-room/statements-releases/2021/04/23/leaders-summit-on-climate-summary-of-proceedings/.

13. E2 Letter to Congress, September 17, 2021, https://e2.org/wp-content/uploads/2021/09/AJP-Merged-Letter-w-Signatures-FINAL.pdf.

14. The White House, "ICYMI: More than 1,000 Business Leaders Call on Congress to Build Back Better on Climate," September 27, 2021, https://e2.org/wp-content/uploads/2021/10/White-House-More-Than-1000-Business-Leaders-Call-on-Congress-to-Build-Back-Better-on-Climate.pdf.

15. We Mean Business Coalition, "More than 330 Major Businesses Call on US Congress To Build Back A More Resilient, Sustainable Economy From Covid-19," May 13, 2021, https://www.wemeanbusinesscoalition.org/press-release/more-than-330-major-businesses-call-on-u-s-congress-to-build-back-a-more-resilient-sustainable-economy-from-covid-19/.

16. U.S. Chamber of Commerce, "US Chamber Statement on Biden Climate Goal," April 22, 2021, https://www.uschamber.com/environment/us-chamber-statement-biden-climate-goal.

17. U.S. Chamber of Commerce, "US Chamber Vows to Defeat Reconciliation, Hails Voting Deadline for Bipartisan Infrastructure Deal," August 24, 2021, https://www.uschamber.com/infrastructure/us-chamber-vows-defeat-reconciliation-hails-voting-deadline-bipartisan-infrastructure.

18. Jonathan Martin and Alexander Burns, "Reeling From Surprise Losses, Democrats Sound the Alarm for 2022," *New York Times*, Nov. 3, 2021, https://www.nytimes.com/2021/11/03/us/politics/democrat-losses-2022.html.

19. Alexi McCammond, "AOC, The Squad defend infrastructure 'no' vote," *Axios*, November 9, 2021, https://www.axios.com/aoc-squad-defend-infrastructure-no-vote-69729c85-10fe-4e71-b0c3-8f99d936626f.html.

20. C-SPAN, "Senator Manchin News Conference on Infrastructure and Spending Negotiations," November 1, 2021, https://www.c-span.org/video/?515788-1/senator-manchin-calls-house-vote-infrastructure-bill-clarity-budget-reconciliation-bill&live.

21. Leah Ann Caldwell and Haley Talbot, "To get the infrastructure bill passed, House Democrats went looking for trust," *NBC News*, November 10, 2021, https://

www.nbcnews.com/politics/congress/get-infrastructure-bill-passed-house-democrats-went-looking-trust-n1283704.

22. Jonathan Weisman and Carl Hulse, "How a $1 Trillion Infrastructure Bill Survived an Intraparty Brawl," *New York Times*, November 6, 2021, https://www.nytimes.com/2021/11/06/us/politics/infrastructure-black-caucus-vote.html.

23. The White House, "Fact Sheet: The Bipartisan Infrastructure Deal," November 6, 2021, https://www.whitehouse.gov/briefing-room/statements-releases/2021/11/06/fact-sheet-the-bipartisan-infrastructure-deal/.

24. The White House, "President Biden Delivers Remarks on the Passage of the Bipartisan Infrastructure Bill," November 6, 2021, https://www.youtube.com/watch?v=baFaMfPjqT0.

25. C-SPAN, "House Minority Leader Kevin McCarthy Concludes Longest Floor Speech," November 18, 2021, https://www.c-span.org/video/?c4987457/house-minority-leader-kevin-mccarthy-concludes-longest-floor-speech.

26. C-SPAN, "House Speaker Pelosi Holds News Conference on Build Back Better Passage," November 19, 2021, https://www.c-span.org/video/?516226-1/speaker-pelosi-passage-social-spending-bill-monumental-historic-transformative.

27. Paul E. Anderson, "Sam Rayburn and Rural Electrification," East Texas History, https://easttexashistory.org/items/show/73.

28. U.S. Department of Transportation, "History of the Interstate Highway System," https://www.fhwa.dot.gov/interstate/history.cfm.

29. NASA, "Apollo 11, July 20, 1969, One Giant Leap for Mankind," July 20, 2019, https://www.nasa.gov/mission_pages/apollo/apollo11.html.

CHAPTER 9

1. Alexander Hamilton, "Report on the Subject of Manufactures," December 5, 1791, National Archives, https://founders.archives.gov/documents/Hamilton/01-10-02-0001-0007.

2. Thomas Jefferson, "From Thomas Jefferson to George Washington," September 9, 1972, National Archives, https://founders.archives.gov/documents/Jefferson/01-24-02-0330.

3. Douglas A. Irwin, "The Aftermath of Hamilton's 'Report on Manufactures,'" *Journal of Economic History,* Vol. 64, No. 3, September 2004, 800–821, https://www.jstor.org/stable/3874820.

4. Searching In History, "Samuel Slater: The Father of American Industrial Revolution," January 16, 2015, https://searchinginhistory.blogspot.com/2015/01/samuel-slater-father-of-american.html.

5. Irwin, "The Aftermath of Hamilton's 'Report on Manufactures.'"

6. Alice Buck, "A History of Energy Research and Development Administration," U.S. Department of Energy, March 1982, https://www.energy.gov/sites/prod/files/ERDA%20History.pdf.

7. Jimmy Carter, "Proclamation 4558—Sun Day 1978," The American Presidency Project, https://www.presidency.ucsb.edu/documents/proclamation-4558-sun-day-1978.

8. U.S. Department of Energy, "Solar Energy in the United States," https://www.energy.gov/eere/solar/solar-energy-united-states.

9. Alan Blinder, "Jimmy Carter Makes a Stand for Solar, Decades after the Cardigan Sweater," *New York Times,* February 11, 2017, https://www.nytimes.com/2017/02/11/us/jimmy-carter-solar-energy-plains-ga.html.

10. Maria Gallucci, "California's Landmark Clean Car Mandate: How It Works and What It Means," *Inside Climate News*, February 8, 2012, https://insideclimatenews.org/news/08022012/california-landmark-clean-vehicles-electric-cars-regulations-2025-fuel-efficiency-epa/.

11. Gary Witzenburg, "GM's EV1 Electric Car Invented Many Technologies That Are Commonplace on Today's EVs," July 7, 2021, *Car and Driver*, https://www.caranddriver.com/news/a36887553/gm-ev1-electric-car-technology/.

12. John O'Dell, "GM Sues to Overturn State's Zero Emission Vehicle Mandate," *Los Angeles Times*, February 24, 2001, https://www.latimes.com/archives/la-xpm-2001-feb-24-fi-29699-story.html.

13. Business Council for Sustainable Energy, "BCSE Webinar: We Have the Technologies and Tools. It's Time to Raise Climate Ambition," September 20, 2021, https://us02web.zoom.us/webinar/register/WN_rfeA1WcCReKk3Mjfhj5m0A.

14. Lazard, "Levelized Cost of Energy and Levelized Cost of Storage, 2019," November 7, 2019, https://www.lazard.com/perspective/lcoe2019.

15. Kelley Blue Book, "Average New-Vehicle Transaction Prices Top $54,000 For First Time," October 13, 2021, https://mediaroom.kbb.com/2021-10-13-Average-New-Vehicle-Transaction-Prices-Top-45,000-for-First-Time,-According-to-Kelley-Blue-Book.

16. Kelley Blue Book, "Electrified Vehicles Sales Boom in Q3," October 26, 2021, https://mediaroom.kbb.com/2021-10-26-Electrified-Vehicle-Sales-Boom-in-Q3-to-Surge-Past-One-Million-Units-in-2021,-According-to-Kelley-Blue-Book-Report.

17. Wesley Cole, A. Will Frazier, and Chad Augustine, "Cost Projections for Utility-Scale Battery Storage: 2021 Update," June 2021, https://www.nrel.gov/docs/fy21osti/79236.pdf.

18. Victoria Tomlinson, "UPS invests in Arrival and orders 10,000 Generation 2 Electric Vehicles," Arrival.com, April 24, 2020, https://arrival.com/us/en/news/ups-invests-in-arrival-and-orders-10000-generation-2-electric-vehicles.

19. Erik Schatzker, "Hertz Order for 100,000 EVs Sends Tesla Value to $1 Trillion," *Bloomberg*, October 25, 2021, https://www.bloomberg.com/news/articles/2021-10-25/hertz-said-to-order-100-000-teslas-in-car-rental-market-shake-up.

20. Al Root, "Ford's Electric F-150 Hits 130,000 Reservations. How That Stacks Up against Other EV Makers," *Barron's*, September 3, 2021, https://www.barrons.com/articles/ford-electric-f-150-reservations-tesla-ev-makers-51630674776.

21. Reuters, "Volvo Group says launches world's first fossil-free steel vehicle," October 13, 2021, https://www.reuters.com/business/sustainable-business/volvo-trucks-says-launches-worlds-first-fossil-free-steel-vehicle-2021-10-13/.

22. Nextera.com, https://www.nexteraenergyresources.com/what-we-do/wind.html.

23. Berkshire Hathaway Energy, "Wind XII Project Positions MidAmerican Energy to Hit 100 Percent Renewable Goal," May 30, 2018, https://www.brkenergy.com/news/wind-xii-project-positions-midamerican-energy-to-hit-100-percent-renewable-goal.

24. Southern Company CEO Tom Fanning on Q32021 Results, November 4, 2021, https://seekingalpha.com/article/4465705-southern-company-ceo-tom-fanning-on-q3-2021-results-earnings-call-transcript.

25. Dominion Energy, "Dominion Energy Submits Application for Coastal Virginia Offshore Wind with Virginia State Corporation Commission," November 5, 2021, https://www.prnewswire.com/news-releases/dominion-energy-submits-application-for-coastal-virginia-offshore-wind-with-virginia-state-corporation-commission-301417477.html.

26. CEBA, "CEBA Announces Top 10 U.S. Large Energy Buyers in 2020," February 10, 2021, https://cebuyers.org/blog/reba-announces-top-10-u-s-large-energy-buyers-in-2020-2/.

27. U.S. EPA, "Green Power Partnership 100% Green Power Users," https://www.epa.gov/greenpower/green-power-partnership-100-green-power-users.

28. McKinsey & Co., "An integrated perspective on the future of mobility, part 2: transforming urban delivery," September 2017, https://www.mckinsey.com/~/media/mckinsey/business%20functions/sustainability/our%20insights/urban%20commercial%20transport%20and%20the%20future%20of%20mobility/an-integrated-perspective-on-the-future-of-mobility.pdf.

29. McKinsey & Co., "Charging electric-vehicle fleets: How to seize the emerging opportunity," March 2020, https://www.mckinsey.com/~/media/McKinsey/Business%20Functions/Sustainability/Our%20Insights/Charging%20electric%20vehicle%20fleets%20How%20to%20seize%20the%20emerging%20opportunity/Charging-electric-vehicle-fleets-how-to-seize-the-emerging-opportunity-FINAL.pdf.

30. U.S. Department of Energy, "DOE Factsheet."

31. E2, "E2 National Webinar: A Conversation with National Climate Advisor Gina McCarthy," September 29, 2021, https://e2.org/events/e2-national-webinar-a-conversation-with-national-climate-advisor-gina-mccarthy/.

CHAPTER 10

1. C-SPAN, "User Clip: President Carter's Fireside Chat on Energy, February 2, 1977," https://www.c-span.org/video/?c4972739/user-clip-president-carters-fireside-chat-energy-feb-2-1977.

2. NOAA Office of Response and Restoration, "1976: A Winter of Ship Accidents," November 10, 2016, https://response.restoration.noaa.gov/oil-and-chemical-spills/significant-incidents/argo-merchant-oil-spill/1976-winter-ship-accidents.htm.

3. Adonis Hoffman, "President Biden, oil and gas power America. Now is not the time to bite the hand of Big Oil," Fox News, February 20, 2021, https://www.foxnews.com/opinion/biden-oil-and-gas-power-america-big-oil-adonis-hoffman.

4. Editorial board, "More Green Blackouts Ahead," *Wall Street Journal*, February 23, 2021, https://www.wsj.com/articles/more-green-blackouts-ahead-11614125061.

5. Mitch McConnell, 'Piecemeal Green New Deal' Putting Ideology Ahead of Working Americans," January 28, 2021, https://www.republicanleader.senate.gov/newsroom/remarks/piecemeal-green-new-deal-putting-ideology-ahead-of-working-americans.

6. Jim Jordan (@Jim_Jordan), "Joe Biden is the new Jimmy Carter," May 11, 2021, Twitter, https://twitter.com/jim_jordan/status/1392099930978803717.

7. Lazard, "Levelized Cost of Energy."

8. *Consumer Reports*, "Electric vehicle owners spending half as much on maintenance compared to gas-powered vehicle owners, finds new CR analysis," September 24, 2020, https://advocacy.consumerreports.org/press_release/electric-vehicle-owners-spending-half-as-much-on-maintenance-compared-to-gas-powered-vehicle-owners-finds-new-cr-analysis/.

9. Benjamin Preston, "Consumer Reports Survey Shows Strong Interest in Electric Cars," *Consumer Reports,* December 18, 2020, https://www.consumerreports.org/hybrids-evs/cr-survey-shows-strong-interest-in-evs-a1481807376/.

10. Brian Kennedy and Alison Spencer, "Most Americans support expanding solar and wind energy, but Republican support has dropped," Pew Research Center, June 8, 2021, https://www.pewresearch.org/fact-tank/2021/06/08/most-americans-support-expanding-solar-and-wind-energy-but-republican-support-has-dropped/.

11. Citizens for Responsible Energy Solutions, "Poll: GOP Voters Overwhelmingly Support Clean Energy to Jumpstart Economy, Create Global Competitive Advantage," June 25, 2020, https://cresenergy.com/pressreleases/poll-gop-voters-overwhelmingly-support-clean-energy-to-jumpstart-economy-create-global-competitive-advantage/.

12. Sarah Kaplan and Andrew Ba Tran, "Nearly 1 in 3 Americans experienced a weather disaster this summer," *Washington Post,* September 4, 2021, https://www.washingtonpost.com/climate-environment/2021/09/04/climate-disaster-hurricane-ida/.

13. E2, "A Conversation with National Climate Advisor Gina McCarthy."

14. Jesse Broehl, "Colorado Voters Pass Renewable Energy Standard," *Renewable Energy World,* November 3, 2004, https://www.renewableenergyworld.com/baseload/colorado-voters-pass-renewable-energy-standard-17736/#gref.

15. Colorado Office of Economic Development & International Trade, "Colorado's Gone in the Wind With Vestas," https://choosecolorado.com/success-stories/vestas/.

16. Associated Press, "Vestas plans 400-employee expansion in Windsor," *Denver Post*, March 13, 2015, https://www.denverpost.com/2015/03/13/vestas-plans-400-employee-expansion-in-windsor/.

17. Judith Kohler, "Gov. Polis signs bills on renewable energy, but what does that mean for Colorado's energy future?" *Denver Post*, May 30, 2019, https://www.denverpost.com/2019/05/30/colorado-jared-polis-renewable-energy-climate/.

18. U.S. Energy Information Administration, "Iowa State Profile and Energy Estimates," June 17, 2021, https://www.eia.gov/state/analysis.php?sid=IA.

19. Chuck Grassley, "Grassley Celebrates Wind Energy Becoming Iowa's Largest Source of Electricity," April 17, 2020, https://www.grassley.senate.gov/news/news-releases/grassley-celebrates-wind-energy-becoming-iowa-s-largest-source-electricity.

20. NC Clean Energy Technology Center, "DSIRE Renewable & Clean Energy Standards," September 2020, https://ncsolarcen-prod.s3.amazonaws.com/wp-content/uploads/2020/09/RPS-CES-Sept2020.pdf.

21. NRDC, "California's Energy Efficiency Success Story," July 2013, https://www.nrdc.org/sites/default/files/ca-success-story-FS.pdf.

22. E2, E4theFuture, "Energy Efficiency Jobs In America," October 2021, https://e2.org/reports/energy-efficiency-jobs-in-america-2021/.

23. Margot Roosevelt, "How Fran Pavley changed the American auto industry," *Los Angeles Times*, April 1, 2010, https://latimesblogs.latimes.com/greenspace/2010/04/fuel-economy-global-warming.html.

24. Interview with Fran Pavley, November 15, 2021.

25. E2, NRDC, "Let's Talk About Clean Jobs: A Conversation with U.S. Secretary of Energy Jennifer Granholm, Governor of Illinois J. B. Pritzker and President and CEO of NRDC Manish Bapna," November 17, 2020.

26. Governor Roy Cooper, "Governor Cooper Signs Energy Bill Including Carbon Reduction Goals into Law," October 13, 2021, https://governor.nc.gov/news/press-releases/2021/10/13/governor-cooper-signs-energy-bill-including-carbon-reduction-goals-law.

27. Illinois.gov, "Gov. Pritzker Signs Transformative Legislation Establishing Illinois as a National Leader on Climate Action," September 15, 2021, https://www.illinois.gov/news/press-release.23893.html.

28. First Solar, "First Solar to Invest $680m in Expanding American Solar Manufacturing Capacity by 3.3 GW," June 9, 2021, https://investor.firstsolar.com/news/press-release-details/2021/First-Solar-to-Invest-680m-in-Expanding-American-Solar-Manufacturing-Capacity-by-3.3-GW/default.aspx.

29. GM, "Wabtec and GM to Develop Advanced Ultium Battery and HYDROTECH Hydrogen Fuel Cell Solutions for Rail Industry," June 15, 2021, https://plants.gm.com/media/us/en/gm/home.detail.html/content/Pages/news/us/en/2021/jun/0615-wabtec.html.

30. Ford, "Ford, SK Innovation add 6,000 Jobs in Tennessee," September 20, 2021, https://media.ford.com/content/fordmedia/fna/us/en/news/2021/09/29/ford--sk-innovation-add-6-000-jobs-in-tennessee.html.

31. Business Network for Offshore Wind, "Statement: New Va. offshore wind blade facility critical step for U.S. supply chain localization," October 23, 2021, https://www.offshorewindus.org/2021/10/25/statement-new-va-offshore-wind-blade-facility-critical-step-for-u-s-supply-chain-localization/.

32. E2, "Clean Jobs Nevada 2019," April 4, 2019, https://e2.org/reports/clean-jobs-nevada-2019/.

33. E2, "CEOs, Execs: Passing Build Back Better Act and Infrastructure Bills Urgent for Economy, Business," October 4, 2021, https://e2.org/releases/ceos-execs-passing-build-back-better-act-now-necessary-to-businesses/.

34. E2, "Clean Jobs America 2021."

35. E2, "Clean Jobs, Better Jobs," October 2020, https://e2.org/wp-content/uploads/2020/10/Clean-Jobs-Better-Jobs.-October-2020.-E2-ACORE-CELI.pdf.

CHAPTER 11

1. GreenBiz.com, "The Climate Capital Cavalry Has Arrived. Now What?" GreenBiz VERGE 2021, November 5, 2021. https://www.greenbiz.com/video/climate-capital-cavalry-has-arrived-now-what.

2. Svenja Telle, "COP26 and the climate finance bubble," *PitchBook*, October 30, 2021, https://pitchbook.com/news/articles/cop26-2021-climate-change-finance-bubble.

3. GreenBiz.com, "The Climate Cavalry Has Arrived."

4. Chris Sacca, "Cash Cools Everything Around Me," Lowercarbon Capital, August 12, 2021, https://lowercarboncapital.com/2021/08/12/cash-cools-everything-around-me/.

5. Lowercarbon Capital, "Kickass Companies," https://lowercarboncapital.com/companies/.

6. Climate Tech VC, "Climate tech $16b mid-year investment action report," August 30, 2021, https://climatetechvc.org/%F0%9F%8C%8F-climate-tech-16b-mid-year-investment-action-report/.

7. E2, "Healthy Soils and the Climate Connection," February 9, 2021, https://e2.org/reports/healthy-soils-climate-connection-2021/.

8. Heart Aerospace, "Heart Aerospace is one step closer to building an electric plane," July 13, 2021, https://heartaerospace.com/wp-content/uploads/2021/07/Heart-Aerospace-Series-A-Press-Release-July-13-2021.pdf.

9. ZeroAvia, "ZeroAvia Secures Additional $24.3 Million," March 31, 2021, https://www.prnewswire.com/news-releases/zeroavia-secures-additional-24-3-million-to-kick-off-large-engine-development-for-50-seat-zero-emission-aircraft-301259265.html.

10. Dr. Benjamin Gaddy, Dr. Varun Sivaram, and Dr. Francis O'Sullivan, "Venture Capital and Cleantech: The Wrong Model for Clean Energy Innovation," MIT, July 2016, https://energy.mit.edu/wp-content/uploads/2016/07/MITEI-WP-2016-06.pdf.

11. Interview with Eric Berman, November 19, 2021.

12. The White House, "Fact Sheet: President Biden Signs Executive Order Catalyzing America's Clean Energy Economy Through Federal Stability," December 8, 2021, https://www.whitehouse.gov/briefing-room/statements-releases/2021/12/08/fact-sheet-president-biden-signs-executive-order-catalyzing-americas-clean-energy-economy-through-federal-sustainability/.

13. Will Wade, "Tom Steyer Forms Fund for Global Fight Against Climate Change," *Bloomberg*, September 9, 2021, https://www.bloomberg.com/news/articles/2021-09-09/tom-steyer-forms-fund-for-global-fight-against-climate-change.

14. Climate Registry, Climate Action Reserve webinar, "Armchair Discussion with National Climate Advisor Gina McCarthy and Galvanize Climate Solutions Co-Founder Tom Steyer," November 8, 2021, https://www.youtube.com/watch?v=vpqRQYm3qDA&list=PLSKKUDlzU1_zoOlOYLAE3X7FRDrLfskYo&index=8.

15. GreenBiz.com, "The Climate Cavalry Has Arrived."

16. Abt Associates webinar, "Perspectives on Public-Private Partnership for Climate Investments," November 5, 2020, https://twitter.com/abtassociates/status/1455187949658775571/photo/1.

CHAPTER 12

1. John Barrasso, "The Biden Administration's Green Energy Priorities Are a Recipe for Repeated Disaster," May 22, 2021, https://www.energy.senate.gov/2021/5/barrasso-the-biden-administration-s-green-energy-priorities-are-a-recipe-for-repeated-disaster.

2. John Broder, "Wyoming Senator Seeks to Lasso E.P.A.," *New York Times,* January 31, 2011, https://green.blogs.nytimes.com/2011/01/31/wyoming-senator-seeks-to-lasso-e-p-a/.

3. Kyle Hyatt, "A Republican senator is trying to kill the federal EV tax credit, again," *CNET Road/Show*, October 10, 2018, https://www.cnet.com/roadshow/news/republican-kill-ev-tax-credit-bill/.

4. John Barrasso, "Barrasso & Rep. Smith Introduce Legislation to End Electric Vehicle Tax Credits," June 9, 2021, https://www.barrasso.senate.gov/public/index.cfm/news-releases?ID=D4A971F0-DC0D-4D90-BD99-826105610B51.

5. Camille Erikson, "Wyoming's rig count falls to zero for only second time in 136 years," *Casper Star-Tribune*, August 4, 2020, https://trib.com/business/energy/wyomings-rig-count-falls-to-zero-for-only-second-time-in-136-years/article_f76c8333-ed38-5c8b-8062-24d4ae1d1c52.html.

6. Matthew Bandyk, "Largest planned wind farm in US gets key federal approval," *UtilityDive,* October 25, 2019, https://www.utilitydive.com/news/largest-planned-wind-farm-in-us-gets-key-federal-approval/565795/.

7. Nicole Pollack, "Six new wind farms proposed for Wyoming," *Casper Star-Tribune*, June 17, 2021, https://trib.com/business/energy/six-new-wind-farms-proposed-for-wyoming/article_41cbf2db-e482-5bb0-a629-fbabdb7296da.html.

8. E2, "Clean Jobs America 2021."

9. OpenSecrets, "Sen. John A Barrasso—Campaign Finance Summary," https://www.opensecrets.org/members-of-congress/summary?cid=N00006236&cycle=CAREER.

10. U.S. Department of Energy, "Electric Vehicle Registrations by State," June 2021, https://afdc.energy.gov/data/10962.

11. Thom Tillis, "Tillis, Colleagues Urge Senate Leaders to Consider Clean Energy Jobs Impacted by Pandemic in Next COVID-19 Package," July 23, 2020, https://www.tillis.senate.gov/2020/7/tillis-colleagues-urge-senate-leaders-to-consider-clean-energy-jobs-impacted-by-pandemic-in-next-covid-19-package.

12. Allan Smith, "McConnell says he's '100 percent' focused on 'stopping' Biden's administration," *NBC News,* May 5, 2021, https://www.nbcnews.com/politics/joe-biden/mcconnell-says-he-s-100-percent-focused-stopping-biden-s-n1266443.

13. E2, "Clean Energy Jobs by Congressional District," November 2021, https://cleanjobsamerica.e2.org/.

14. Matt Canham, "Look who's leading the charge for a GOP climate-change caucus: a Utah Republican," *Salt Lake Tribune*, June 11, 2021, https://www.sltrib.com/news/politics/2021/06/11/look-whos-leading-charge/.

15. American Conservation Coalition, "Poll, Youth and Climate Change," February 10, 2020, https://www.acc.eco/acc-youthpoll.

16. Bethany Bowra, "Press Release: ACC Releases National Millennial and Gen-Z Polling on Climate Change and Potential Solutions," American Conservation Coalition, February 10, 2020, https://www.acc.eco/blog/2020/2/10/press-release-acc-releases-national-millennial-and-gen-z-polling-on-climate-change-and-potential-solutions.

17. Interview with Mark Fleming, October 25, 2021.

18. John F. Kennedy, "Address to Joint Session of Congress May 25, 1961," John F. Kennedy Presidential Library and Museum, https://www.jfklibrary.org/learn/about-jfk/historic-speeches/address-to-joint-session-of-congress-may-25-1961.

19. NASA, "Fact Sheets NASA Langley Research Center's Contribution to the Apollo Program," https://www.nasa.gov/centers/langley/news/factsheets/Apollo.html.

20. U.S. Department of Energy, "Remarks as delivered by Secretary Granholm at President Biden's Leaders Summit on Climate," April 23, 2021, https://www.energy.gov/articles/remarks-delivered-secretary-granholm-president-bidens-leaders-summit-climate.

CHAPTER 13

1. United Nations Climate Change, "Speech, Mia Mottley, Prime Minister of Barbados at the Opening of the Cop26 World Leaders Summit," November 1, 2021, https://www.youtube.com/watch?v=PN6THYZ4ngM.

2. UN Climate Change Conference, "COP26 Goals," https://ukcop26.org/cop26-goals/.

3. Emily Beament, "Greater sense of urgency at COP26 but 'job not done'—John Kerry," *The Standard*, November 5, 2021, https://www.standard.co.uk/news/uk/john-kerry-cop26-joe-biden-president-glasgow-b964666.html.

4. Simon Jessop, Katy Daigle, and Valerie Volcovici, "Baffled? You're not alone—COP26 spawns confusing array of acronyms," *Reuters*, November 5, 2012,

https://www.reuters.com/business/cop/mission-coalition-cop26-spawns-confusing-clusters-2021-11-05/.

5. Bill Gates, "In Glasgow I saw three big shifts in the climate conversation," *GatesNotes*, November 8, 2021, https://www.gatesnotes.com/Energy/Reflections-from-COP26.

6. Joel Makower, "The business of COP is business," *GreenBiz*, November 8, 2021, https://www.greenbiz.com/article/business-cop-business.

7. Rockefeller Foundation, "Historic Alliance Launches at COP26 to Accelerate a Transition to Renewable Energy, Access to Energy for All, and Jobs," November 1, 2021, https://www.rockefellerfoundation.org/news/historic-alliance-launches-at-cop26-to-accelerate-renewable-energy-climate-solutions-and-jobs/.

8. Glasgow Financial Alliance for Net Zero, "Amount of finance committed to achieving 1.5C now at scale needed to deliver the transition," November 3, 2021, https://www.gfanzero.com/press/amount-of-finance-committed-to-achieving-1-5c-now-at-scale-needed-to-deliver-the-transition/.

9. PPCA, "New PPCA members tip the scales toward 'consigning coal to history,' at COP26," November 4, 2021, https://www.poweringpastcoal.org/news/press-release/new-ppca-members-tip-the-scales-towards-consigning-coal-to-history-at-cop26.

10. Global Cement and Concrete Association, "Global Cement and Concrete Industry Announces Roadmap to Achieve Groundbreaking 'Net Zero' CO2 Emissions by 2050," October 12, 2021, https://gccassociation.org/news/global-cement-and-concrete-industry-announces-roadmap-to-achieve-groundbreaking-net-zero-co2-emissions-by-2050/.

11. Steven Mufson, "What could finally stop new coal plants? Pulling the plug on their insurance," *Washington Post,* October 26, 2021, https://www.washingtonpost.com/climate-environment/2021/10/26/climate-change-insurance-coal/.

12. Climate Investment Funds, "New CIF Bond Offering Could Mobilize $50B for Clean Technologies in Developing Countries," November 3, 2021, https://www.climateinvestmentfunds.org/news/new-cif-bond-offering-could-mobilize-50b-clean-technologies-developing-countries.

13. U.S. Department of the Treasury, "Remarks by Secretary of the Treasury Janet L. Yellen at COP 26 in Glasgow, Scotland," November 3, 2021, https://home.treasury.gov/news/press-releases/jy0465.

14. USAID, "How the United States Benefits from Agricultural and Food Security Investments in Developing Countries," https://www.usaid.gov/sites/default/files/documents/1867/BIFAD_US_Benefit_Study.pdf

15. U.S. Foreign Agricultural Service, "Soybeans," https://www.fas.usda.gov/commodities/soybeans.

16. Clyde Russell, "Coal trajectory is set whether it's 'phase out' or 'phase down,'" *Reuters*, November 15, 2021, https://www.reuters.com/business/cop/coal-trajectory-is-set-whether-its-phase-out-or-phase-down-russell-2021-11-14/.

17. *Wall Street Journal*, "Kerry Addresses 'Phase Down' of Coal in COP26 Agreement," November 13, 2021, https://www.wsj.com/video/kerry-addresses-phase-down-of-coal-in-cop26-agreement/2803D978-DEDB-41B1-93CE-BD01D450A8D8.html.

18. *Ahram Online*, "Scorpion stings kill 3, injure 450 amid bad weather in Egypt's Aswan," November 13, 2021, https://english.ahram.org.eg/NewsContent/1/64/439572/Egypt/Politics-/Scorpion-stings-kill-,-injure-amid-bad-weather-in.aspx.

19. 4NewYork, "It's Devastating: First-Ever November Tornadoes Leave LI Families Floundering," November 15, 2021, https://www.nbcnewyork.com/news/local/clean-up-continues-after-rare-november-tornadoes-wreak-havoc-on-long-island-ny-only/3401674/.

CHAPTER 14

1. Ian Morse, "Down on the Farm That Harvests Metal from Plants," *New York Times*, February 26, 2020, https://www.nytimes.com/2020/02/26/science/metal-plants-farm.html.

2. Antony van der Ent et al., "Nickel biopathways in tropical nickel hyperaccumulating trees from Sabah (Malaysia)," *Nature.com*, February 16, 2017, https://www.nature.com/articles/srep41861.

3. Don Comis, "There's Metals in Them Thar Plants!" USDA Agricultural Research Service, June 22, 2000, https://www.ars.usda.gov/news-events/news/research-news/2000/theres-metals-in-them-thar-plants/.

4. Daniel Malloy, "Johnny Isakson tussles with a Democrat on Senate Floor: 'I don't believe climate change is a religion,'" *Atlanta Journal-Constitution*, March 18, 2015, https://www.ajc.com/blog/politics/johnny-isakson-tussles-with-democrat-senate-floor-don-believe-climate-change-religion/ajruIc3fbNCxmIHFNplS8J/.

5. Paul Kane, "A beloved senator heads into retirement with no clear successor to fill the void," *Washington Post*, December 5, 2019, https://www.washingtonpost.com/powerpost/a-beloved-senator-heads-into-retirement-with-no-clear-successor-to-fill-the-void/2019/12/05/6922d9b6-16b4-11ea-9110-3b34ce1d92b1_story.html.

6. Theodore Roosevelt, "Conservation as a National Duty," *Voices of Democracy*, May 13, 1908, http://voicesofdemocracy.umd.edu/theodore-roosevelt-conservation-as-a-national-duty-speech-text/.

7. Douglas Brinkley, *The Wilderness Warrior: Theodore Roosevelt and the Crusade for America*, New York: HarperCollins, 2009.

Bibliography

4NewYork. (2021). "It's Devastating: First-Ever November Tornadoes Leave LI Families Floundering." https://www.nbcnewyork.com/news/local/clean-up-continues-after-rare-november-tornadoes-wreak-havoc-on-long-island-ny-only/3401674/.

Abt Associates, webinar. (2020). "Perspectives on Public-Private Partnership for Climate Investments." https://twitter.com/abtassociates/status/1455187949658775571/photo/1.

Adams, Dwight. (2018). "RFRA: Why the 'religious freedom law' signed by Mike Pence was so controversial." *IndyStar*. https://www.indystar.com/story/news/2018/04/25/rfra-indiana-why-law-signed-mike-pence-so-controversial/546411002/.

Ahram Online. (2021). "Scorpion stings kill 3, injure 450 amid bad weather in Egypt's Aswan." https://english.ahram.org.eg/NewsContent/1/64/439572/Egypt/Politics-/Scorpion-stings-kill-,-injure-amid-bad-weather-in.aspx.

Amazon. (2021). "Amazon is making big global investments in renewable energy." https://www.aboutamazon.com/news/sustainability/amazon-is-making-big-global-investments-in-renewable-energy.

American Conservation Coalition. (2020). "Poll, Youth and Climate Change." https://www.acc.eco/acc-youthpoll.

American Farm Bureau Federation. (2019). "Farm Bankruptcies Rise Again." https://www.fb.org/market-intel/farm-bankruptcies-rise-again.

Anderson, Paul E. (2022). "Sam Rayburn and Rural Electrification." East Texas History. https://easttexashistory.org/items/show/73.

Andrews, Travis M. (2017). "Kentucky Coal Mining Museum in Harlan County switches to solar power." *Washington Post*. https://www.washingtonpost.com/news/morning-mix/wp/2017/04/06/the-coal-mining-museum-in-harlan-county-ky-switches-to-solar-power/.

Argus Research. (2017). "US farm debt declines on high crop prices, federal aid." https://www.argusmedia.com/en/news/2232923-us-farm-debt-declines-on-high-crop-prices-federal-aid.

As You Sow. (2021). "92% of Sysco Shareholders Support Net-Zero Climate Targets." https://www.asyousow.org/press-releases/2021/11/29/sysco-shareholders-support-net-zero-climate-targets.

Associated Press. (2015). "Vestas plans 400-employee expansion in Windsor." *Denver Post*. https://www.denverpost.com/2015/03/13/vestas-plans-400-employee-expansion-in-windsor/.

Associated Press. (2016). "Radioactive materials from nuclear plant leaking into Biscayne Bay, study says." *Orlando Sentinel*. https://www.orlandosentinel.com/features/gone-viral/os-turkey-point-nuclear-radioactive-biscayne-bay-story.html.

Associated Press. (2017). "'Bathroom bill' to cost North Carolina $3.76 billion." *CNBC*. https://www.cnbc.com/2017/03/27/bathroom-bill-to-cost-north-carolina-376-billion.html.

Associated Press. (2018). "7,000 working at Tesla Gigafactory east of Reno." *Las Vegas Sun*. https://lasvegassun.com/news/2018/dec/05/7000-working-at-tesla-gigafactory-east-of-reno/.

Baer, Justin, and Dawn Lim. (2021). "The Hedge-Fund Manager Who Did Battle with Exxon—and Won." *Wall Street Journal*. https://www.wsj.com/articles/the-hedge-fund-manager-who-did-battle-with-exxonand-won-11623470420.

Bandyk, Matthew. (2019). "Largest planned wind farm in US gets key federal approval." *UtilityDive*. https://www.utilitydive.com/news/largest-planned-wind-farm-in-us-gets-key-federal-approval/565795/.

Banerjee, Neela, John Cushman, Jr., David Hasemyer, and Lisa Song. (2016). "Exxon: The Road Not Taken." *Inside Climate News*. https://insideclimatenews.org/project/exxon-the-road-not-taken/.

Barrasso, John. (2021). "The Biden Administration's Green Energy Priorities Are a Recipe for Repeated Disaster." https://www.energy.senate.gov/2021/5/barrasso-the-biden-administration-s-green-energy-priorities-are-a-recipe-for-repeated-disaster.

Barrasso, John. (2021). "Barrasso & Rep. Smith Introduce Legislation to End Electric Vehicle Tax Credits." https://www.barrasso.senate.gov/public/index.cfm/news-releases?ID=D4A971F0-DC0D-4D90-BD99-826105610B51.

Barrow, Bill, and Zeke Miller. (2021). "Biden and Carter, longtime allies, reconnect in Georgia." *Associated Press*. https://apnews.com/article/politics-georgia-health-coronavirus-voting-rights-0940cfedf2e7d58612ff58f4ca9a04f4.

Beament, Emily. (2021). "Greater sense of urgency at COP26 but 'job not done'—John Kerry" *The Standard*. https://www.standard.co.uk/news/uk/john-kerry-cop26-joe-biden-president-glasgow-b964666.html.

Benioff, Marc, and Jane Goodall. (2020). "The Trillion Tree challenge that could save the planet." World Economic Forum. https://www.weforum.org/agenda/2020/09/planting-1-trillion-trees-save-the-planet-sustainability/.

Berkshire Hathaway Energy. (2018). "Wind XII Project Positions MidAmerican Energy to Hit 100 Percent Renewable Goal." https://www.brkenergy.com/news/wind-xii-project-positions-midamerican-energy-to-hit-100-percent-renewable-goal.

Berman, Eric. (2021). Interview.

Bermel, Colby. (2020). "Newsom: No patience for climate change deniers amid historic wildfires." *Politico*. https://www.politico.com/states/california/story/2020/09/08/newsom-no-patience-for-climate-deniers-amid-historic-heat-fires-1316014.

BlackRock. (2021). "Larry Fink's 2021 letter to CEOs." https://www.blackrock.com/corporate/investor-relations/larry-fink-ceo-letter.

BlackRock. (2021). "Vote Bulletin: ExxonMobil Corporation." https://www.blackrock.com/corporate/literature/press-release/blk-vote-bulletin-exxon-may-2021.pdf.

Blinder, Alan. (2017). "Jimmy Carter Makes a Stand for Solar, Decades after the Cardigan Sweater." *New York Times*. https://www.nytimes.com/2017/02/11/us/jimmy-carter-solar-energy-plains-ga.html.

Bowra, Bethany. (2020). "Press Release: ACC Releases National Millennial and Gen-Z Polling on Climate Change and Potential Solutions." American Conservation Coalition. https://www.acc.eco/blog/2020/2/10/press-release-acc-releases-national-millennial-and-gen-z-polling-on-climate-change-and-potential-solutions.

Brackett, Ron, and Jan Wesner Childs. (2020). "Tennessee Tornado Death Toll Rises; Several People Still Missing in Putnam County." *Weather.com*. https://weather.com/news/news/2020-03-03-tennessee-tornado-damage-deaths-severe-storms.

Brinkley, Douglas. (2009). *The Wilderness Warrior: Theodore Roosevelt and the Crusade for America*. New York: HarperCollins.

Broder, John. (2011). "Wyoming Senator Seeks to Lasso E.P.A." *New York Times*. https://green.blogs.nytimes.com/2011/01/31/wyoming-senator-seeks-to-lasso-e-p-a/.

Broehl, Jesse. (2004). "Colorado Voters Pass Renewable Energy Standard." *Renewable Energy World*. https://www.renewableenergyworld.com/baseload/colorado-voters-pass-renewable-energy-standard-17736/#gref.

Buck, Alice. (1982). "A History of Energy Research and Development Administration." U.S. Department of Energy. https://www.energy.gov/sites/prod/files/ERDA%20History.pdf.

Business Council for Sustainable Energy. (2021). "BCSE Webinar: We Have the Technologies and Tools. It's Time to Raise Climate Ambition." https://us02web.zoom.us/webinar/register/WN_rfeA1WcCReKk3Mjfhj5m0A.

Business Network for Offshore Wind. (2021). "Statement: New Va. offshore wind blade facility critical step for U.S. supply chain localization." https://www.offshorewindus.org/2021/10/25/statement-new-va-offshore-wind-blade-facility-critical-step-for-u-s-supply-chain-localization/.

Butzer, Stephanie. (2020). "How are Colorado's wildfires still burning in the snow?" *Thedenverchannel.com*. https://www.thedenverchannel.com/news/wildfire/how-are-colorados-wildfires-still-burning-in-the-snow.

Byrne, Kevin. (2020). "Isaias created a tornado outbreak as it raced up the East Coast." *Accuweather*. https://www.accuweather.com/en/hurricane/isaias-created-a-tornado-outbreak-as-it-raced-up-the-east-coast/791095.

Caldwell, Leah Ann, and Haley Talbot. (2021). "To get the infrastructure bill passed, House Democrats went looking for trust." *NBC News*. https://www.nbcnews.com/politics/congress/get-infrastructure-bill-passed-house-democrats-went-looking-trust-n1283704.

California Department of Insurance. (2020). "Commissioner Lara puts focus on solutions." https://www.insurance.ca.gov/0400-news/0100-press-releases/2020/release104-2020.cfm.

Canham, Matt. (2021). "Look who's leading the charge for a GOP climate-change caucus: a Utah Republican." *Salt Lake Tribune*. https://www.sltrib.com/news/politics/2021/06/11/look-whos-leading-charge/.

Capital Dynamics and Switch. (2020). "Switch and Capital Dynamics Break Ground on Massive Solar and Battery Storage Developments." https://www.prnewswire.com/news-releases/switch-and-cital-dynamics-break-ground-on-massive-solar-and-battery-storage-developments-advancing-rob-roys-gigawatt-nevada-301097874.html.

Cappuchi, Matthew. (2021). "Denver still hasn't seen any snow this fall. That's a record." *Washington Post*. https://www.washingtonpost.com/weather/2021/11/22/denver-snow-record-dry/.

Carter, Jimmy. (1978). "Proclamation 4558—Sun Day 1978." The American Presidency Project. https://www.presidency.ucsb.edu/documents/proclamation-4558-sun-day-1978.

CEBA. (2021). "CEBA Announces Top 10 U.S. Large Energy Buyers in 2020." https://cebuyers.org/blog/reba-announces-top-10-u-s-large-energy-buyers-in-2020-2/.

Center for American Progress. (2019). "Trump Administration Is Selling Western Wildlife Corridors to Oil and Gas industry." https://americanprogress.org/article/trump-administration-selling-western-wildlife-corridors-oil-gas-industry/.

CERES. (2021). "411 Businesses and Investors Support U.S. Federal Climate Target in Open Letter to President Biden." https://www.ceres.org/news-center/press-releases/411-businesses-and-investors-support-us-federal-climate-target-open.

CERES. (2021). "Investors seek greater climate action in 2021 Proxy Season." https://www.ceres.org/news-center/press-releases/investors-seek-greater-climate-action-2021-proxy-season.

Charles, Dan. (2020). "Farmers Got a Government Bailout in 2020, Even Those Who Didn't Need It." *NPR*. https://www.npr.org/2020/12/30/949329557/farmers-got-a-government-bailout-in-2020-even-those-who-didnt-need-it.

Chevron. (2021). "2021 Proxy Statement." https://www.chevron.com/-/media/shared-media/documents/chevron-proxy-statement-2021.pdf.

Citizens for Responsible Energy Solutions. (2020). "Poll: GOP Voters Overwhelmingly Support Clean Energy to Jumpstart Economy, Create Global Competitive Advantage." https://cresenergy.com/pressreleases/poll-gop-voters-overwhelmingly-support-clean-energy-to-jumpstart-economy-create-global-competitive-advantage/.

Climate Investment Funds. (2021). "New CIF Bond Offering Could Mobilize $50B for Clean Technologies in Developing Countries." https://www.climateinvestmentfunds.org/news/new-cif-bond-offering-could-mobilize-50b-clean-technologies-developing-countries.

Climate Registry. (2021). Climate Action Reserve webinar, "Armchair Discussion with National Climate Advisor Gina McCarthy and Galvanize Climate Solutions Co-Founder Tom Steyer." https://www.youtube.com/watch?v=vpqRQYm3qDA&list=PLSKKUDlzU1_zoOlOYLAE3X7FRDrLfskYo&index=8.

Climate Tech VC. (2021). "Climate tech $16b mid-year investment action report." https://climatetechvc.org/%F0%9F%8C%8F-climate-tech-16b-mid-year-investment-action-report/.

Climate Voice. (2021). "It's Go Time." https://climatevoice.org/campaigns/.

Clinton White House Archives. (1998). "Testimony of Dr. Janet Yellen, Chair, Council of Economic Advisers." https://clintonwhitehouse2.archives.gov/WH/EOP/CEA/html/19980304.html.

Coalition for Green Capital. (2021). "Clean Energy Accelerator." https://coalitionforgreencapital.com/accelerator/.

Cole, Wesley, A. Will Frazier, and Chad Augustine. (2021). "Cost Projections for Utility-Scale Battery Storage: 2021 Update." https://www.nrel.gov/docs/fy21osti/79236.pdf.

Colorado Office of Economic Development & International Trade. (2021). "Colorado's Gone in the Wind with Vestas." https://choosecolorado.com/success-stories/vestas/.

Comis, Don. (2000). "There's Metals in Them Thar Plants!" USDA Agricultural Research Service. https://www.ars.usda.gov/news-events/news/research-news/2000/theres-metals-in-them-thar-plants/.

Companiesmarketcap.com. (2021). "Largest American companies by market capitalization." https://companiesmarketcap.com/usa/largest-companies-in-the-usa-by-market-cap/.

Congress.gov. (2010). "S.1733 – Clean Energy Jobs and American Power Act." https://www.congress.gov/bill/111th-congress/senate-bill/1733.

Congressional Research Service. (2020). "The Renewable Electricity Production Tax Credit: In Brief." https://sgp.fas.org/crs/misc/R43453.pdf.

Congressional Research Service. (2021). "The Clean Electricity Performance Program (CEPP): In Brief." https://sgp.fas.org/crs/misc/R46934.pdf.

Congressional Research Service. (2021). "The Energy Credit or Energy Investment Tax Credit (ITC)." https://crsreports.congress.gov/product/pdf/IF/IF10479.

Consumer Reports. (2020). "Electric vehicle owners spending half as much on maintenance compared to gas-powered vehicle owners, finds new CR analysis." *Consumer Reports*. https://advocacy.consumerreports.org/press_release/electric-vehicle-owners-spending-half-as-much-on-maintenance-compared-to-gas-powered-vehicle-owners-finds-new-cr-analysis/.

Cornell, Maraya. (2018). "How catastrophic fires have ravaged through California." *National Geographic*. https://www.nationalgeographic.com/environment/article/how-california-fire-catastrophe-unfolded.

Council of Insurance Agents and Brokers. (2021). "Q3 P/C Market Survey 2021." https://www.ciab.com/resources/q3-pc-market-survey-2021/.

Coutu, Peter. (2018). "Could flooding and sea level rise cost Hampton Roads a military base?" *Virginian-Pilot*. https://www.pilotonline.com/military/article_f2bc6da2-975a-11e8-8119-4368a84d4813.html.

Crane-Droesch, Andrew, et al. (2019). "Climate Change Projected To Increase Cost of the Federal Crop Insurance Program." USDA. https://www.ers.usda.gov/amber

-waves/2019/november/climate-change-projected-to-increase-cost-of-the-federal-crop-insurance-program-due-to-greater-insured-value-and-yield-variability/.

C-SPAN. (1997). "User Clip: President Carter's Fireside Chat on Energy, Feb. 2, 1977." https://www.c-span.org/video/?c4972739/user-clip-president-carters-fireside-chat-energy-feb-2-1977.

C-SPAN. (2009). "White House Jobs Summit Opening." https://www.c-span.org/video/?290419-1/white-house-jobs-summit-opening.

C-SPAN. (2019). "House Natural Resources Committee Hearing on Climate Change." https://www.c-span.org/video/?457612-1/house-natural-resources-committee-hearing-climate-change

C-SPAN. (2021). "House Minority Leader Kevin McCarthy Concludes Longest Floor Speech." https://www.c-span.org/video/?c4987457/house-minority-leader-kevin-mccarthy-concludes-longest-floor-speech.

C-SPAN. (2021). "House Speaker Pelosi Holds News Conference on Build Back Better Passage." https://www.c-span.org/video/?516226-1/speaker-pelosi-passage-social-spending-bill-monumental-historic-transformative.

C-SPAN. (2021). "Senator Manchin News Conference on Infrastructure and Spending Negotiations." https://www.c-span.org/video/?515788-1/senator-manchin-calls-house-vote-infrastructure-bill-clarity-budget-reconciliation-bill&live.

Czigler, Thomas, et al. (2020). "Laying the foundation for zero-carbon cement." McKinsey & Co. https://www.mckinsey.com/industries/chemicals/our-insights/laying-the-foundation-for-zero-carbon-cement.

Damian, Johnson. (2021). "10 states with highest homeowners' insurance rates." *Mortgage Professional America*. https://www.mpamag.com/us/mortgage-industry/guides/10-states-with-the-highest-homeowners-insurance-rates/260722.

Davenport, Coral. (2021). "Key to Biden's Climate Agenda Likely to Be Cut Because of Manchin Opposition." *New York Times*. https://www.nytimes.com/2021/10/15/climate/biden-clean-energy-manchin.html.

Delta. (2020). "Delta commits $1 billion to become first carbon neutral airline globally." https://news.delta.com/delta-commits-1-billion-become-first-carbon-neutral-airline-globally.

Department of Defense. (2014). "2014 Climate Change Adaptation Roadmap." https://www.acq.osd.mil/eie/downloads/CCARprint_wForward_e.pdf.

Dietrich, Tamara. (2018). "Study: Sea level rise has already cost Virginians $280 million in home values since 2005." *Daily Press*. https://www.dailypress.com/news/dp-nws-home-value-loss-20180910-story.html.

Dominion Energy. (2021). "Dominion Energy Submits Application for Coastal Virginia Offshore Wind with Virginia State Corporation Commission." https://www.prnewswire.com/news-releases/dominion-energy-submits-application-for-coastal-virginia-offshore-wind-with-virginia-state-corporation-commission-301417477.html.

Doubek, James. (2019). "Air Force Needs Almost $5 Billion to Recover Bases From Hurricane, Flood Damage." *NPR*. https://www.npr.org/2019/03/28/707506544/air-force-needs-almost-5-billion-to-recover-bases-from-hurricane-flood-damage.

Durkin, Erin. (2018). "North Carolina didn't like science on sea levels . . . so it passed a law against it." *The Guardian*. https://www.theguardian.com/us-news/2018/sep/12/north-carolina-didnt-like-science-on-sea-levels-so-passed-a-law-against-it.

E2. (2019). "Clean Jobs Nevada 2019." https://e2.org/reports/clean-jobs-nevada-2019/.

E2. (2021). "CEOs, Execs: Passing Build Back Better Act and Infrastructure Bills Urgent for Economy, Business." https://e2.org/releases/ceos-execs-passing-build-back-better-act-now-necessary-to-businesses/.

E2. (2021). "Clean Energy Jobs by Congressional District." https://cleanjobsamerica.e2.org/.

E2. (2021). "Clean Jobs America 2020." https://e2.org/reports/clean-jobs-america-2020/.

E2. (2021). "Clean Jobs America 2021." https://e2.org/reports/clean-jobs-america-2021/.

E2. (2021). "Healthy Soils and the Climate Connection." https://e2.org/reports/healthy-soils-climate-connection-2021/.

E2. (2021). Letter to Congress. https://e2.org/wp-content/uploads/2021/09/AJP-Merged-Letter-w-Signatures-FINAL.pdf.

E2. (2021). "National Webinar: A Conversation with National Climate Advisor Gina McCarthy." https://e2.org/events/e2-national-webinar-a-conversation-with-national-climate-advisor-gina-mccarthy/.

E2. (2021). "Nearly 400 Business Leaders to Biden: Cut Emissions 50% by 2030." https://e2.org/releases/nearly-400-business-leaders-to-biden-cut-emissions-50-by-2030/.

E2, AABE, Alliance to Save Energy, BOSS, BW Research, and Energy Efficiency for All. (2021). "Help Wanted: Diversity in Clean Energy." https://e2.org/reports/diversity-in-clean-energy-2021/.

E2, ACORE, BW Research, and CELI. (2020). "Clean Jobs, Better Jobs." https://e2.org/wp-content/uploads/2020/10/Clean-Jobs-Better-Jobs.-October-2020.-E2-ACORE-CELI.pdf.

E2, E4theFuture. (2021). "Energy Efficiency Jobs in America." https://e2.org/reports/energy-efficiency-jobs-in-america-2021/.

E2, NRDC. (2020). "Let's Talk About Clean Jobs: A Conversation with U.S. Secretary of Energy Jennifer Granholm, Governor of Illinois J. B. Pritzker and President and CEO of NRDC Manish Bapna."

Energy Optimizers. (2019). "Monroe Local Schools." https://energyoptusa.com/fccasestudies/monroe-local-schools/.

EPA. (2021). "Sources of Greenhouse Gas Emissions." https://www.epa.gov/ghg emissions/sources-greenhouse-gas-emissions.

Erikson, Camille. (2020). "Wyoming's rig count falls to zero for only second time in 136 years." *Casper Star-Tribune*. https://trib.com/business/energy/wyomings-rig-count-falls-to-zero-for-only-second-time-in-136-years/article_f76c8333-ed38-5c8b-8062-24d4ae1d1c52.html.

Exxon Mobil Corp. (2021). Schedule 14A Proxy Statement. https://www.sec.gov/Archives/edgar/data/34088/000090266421001694/p21-0854prec14a.htm.

Fan, Ariel. (2021). Interview.

Federal Register. (2017). "Promoting Energy Independence and Economic Growth." https://www.federalregister.gov/documents/2017/03/31/2017-06576/promoting-energy-independence-and-economic-growth.

Fink, Larry. (2020). "A Fundamental Reshaping of Finance." BlackRock. https://www.blackrock.com/corporate/investor-relations/2020-larry-fink-ceo-letter.

Finney, Michael, and Randall Yip. (2020). "California wildfires could lead to major spikes in cost of home insurance." *ABC 7 News*. https://abc7news.com/home-insurance-fire-fair-plan-risk-score/6797448/.

First Solar. (2021). "First Solar to Invest $680m in Expanding American Solar Manufacturing Capacity by 3.3 GW." https://investor.firstsolar.com/news/press-release-details/2021/First-Solar-to-Invest-680m-in-Expanding-American-Solar-Manufacturing-Capacity-by-3.3-GW/default.aspx.

Fishel, Justin. (2016). "John Kerry Signs Climate Deal with Granddaughter Seated In Lap." *ABC News*. https://abcnews.go.com/International/john-kerry-signs-climate-deal-granddaughter-seated-lap/story?id=38600097.

Fitzpatrick, Jack. (2017). "Interior's Pick to Lead Offshore Oil Regulator Has Industry Ties." *Morning Consult*. https://morningconsult.com/2017/05/22/interiors-pick-lead-offshore-oil-regulator-industry-ties/.

Fleming, Mark. (2021). Interview.

Forbes. (2022). "Janet Yellen Profile." https://www.forbes.com/profile/janet-yellen/?sh=4f96799b36d7.

Ford Motor Co. (2021). "Ford to Lead America's Shift to Electric Vehicles with New Mega Campus." https://media.ford.com/content/fordmedia/fna/us/en/news/2021/09/27/ford-to-lead-americas-shift-to-electric-vehicles.html.

Ford. (2021). "Ford, SK Innovation add 6,000 Jobs in Tennessee." https://media.ford.com/content/fordmedia/fna/us/en/news/2021/09/29/ford--sk-innovation-add-6-000-jobs-in-tennessee.html.

Frazee, Gretchen. (2018). "How climate change is changing your insurance." *PBS Newshour*. https://www.pbs.org/newshour/economy/making-sense/how-climate-change-is-changing-your-insurance.

Gaddy, Benjamin, Varun Sivaram, and Francis O'Sullivan. (2016). "Venture Capital and Cleantech: The Wrong Model for Clean Energy Innovation." MIT. https://energy.mit.edu/wp-content/uploads/2016/07/MITEI-WP-2016-06.pdf.

Gallucci, Maria. (2012). "California's Landmark Clean Car Mandate: How It Works and What It Means." *Inside Climate News*. https://insideclimatenews.org/news/08022012/california-landmark-clean-vehicles-electric-cars-regulations-2025-fuel-efficiency-epa/.

Garamone, Jim. (2014). "Military Must Be Ready For Climate Change, Hagel Says." *DOD News*. https://www.defense.gov/News/News-Stories/Article/Article/603441/military-must-be-ready-for-climate-change-hagel-says/.

Gates, Bill. (2021). "In Glasgow I saw three big shifts in the climate conversation." *GatesNotes*. https://www.gatesnotes.com/Energy/Reflections-from-COP26.

Gelles, David. (2015). "Microsoft Leads Movement to Offset Emissions with Internal Carbon Tax." *New York Times*. https://www.nytimes.com/2015/09/27/business

/energy-environment/microsoft-leads-movement-to-offset-emissions-with-internal-carbon-tax.html.

Gillis, Elsa. (2018). "Hurricane Florence aftermath: 1-year-old dies after vehicle flooded by rising water." *Atlanta Journal-Constitution.* https://www.ajc.com/weather/hurricane-florence-aftermath-year-old-missing-after-vehicle-flooded-rising-waters/Y9SAFhpnshVgmF5t9xSiaM/.

Glasgow Financial Alliance for Net Zero. (2012). "Amount of finance committed to achieving 1.5C now at scale needed to deliver the transition." https://www.gfanzero.com/press/amount-of-finance-committed-to-achieving-1-5c-now-at-scale-needed-to-deliver-the-transition/.

Global Cement and Concrete Association. (2012). "Global Cement and Concrete Industry Announces Roadmap to Achieve Groundbreaking 'Net Zero' CO2 Emissions by 2050." https://gccassociation.org/news/global-cement-and-concrete-industry-announces-roadmap-to-achieve-groundbreaking-net-zero-co2-emissions-by-2050/.

GM. (2012). "Wabtec and GM to Develop Advanced Ultium Battery and HYDROTECH Hydrogen Fuel Cell Solutions for Rail Industry." https://plants.gm.com/media/us/en/gm/home.detail.html/content/Pages/news/us/en/2021/jun/0615-wabtec.html.

Gough, Matt. (2021). "California's Cities Lead the Way to a Gas-Free Future." Sierra Club. https://www.sierraclub.org/articles/2021/07/californias-cities-lead-way-gas-free-future.

Granholm, Jennifer (@JenGranholm). (2021). "My sincere thanks." Twitter. https://twitter.com/jengranholm/status/1365001115893309441?lang=en.

Grassley, Chuck. (2020). "Grassley Celebrates Wind Energy Becoming Iowa's Largest Source of Electricity." https://www.grassley.senate.gov/news/news-releases/grassley-celebrates-wind-energy-becoming-iowa-s-largest-source-electricity.

Greenberg, Jon. (2019). "Donald Trump's ridiculous link between cancer, wind turbines." *Politifact.* https://www.politifact.com/factchecks/2019/apr/08/donald-trump/republicans-dismiss-trumps-windmill-and-cancer-cla/.

GreenBiz.com. (2021). "The Climate Capital Cavalry Has Arrived. Now What?" GreenBiz VERGE 2021. https://www.greenbiz.com/video/climate-capital-cavalry-has-arrived-now-what.

Hamilton, Alexander. (1791). "Report on the Subject of Manufactures." National Archives. https://founders.archives.gov/documents/Hamilton/01-10-02-0001-0007.

Harkins, Gina. (2018). "Lejeune Commander Fires Back at Critics After Declining to Evacuate Base." *Military.com.* https://www.military.com/daily-news/2018/09/12/lejeune-commander-fires-back-critics-after-declining-evacuate-base.html.

Heart Aerospace. (2021). "Heart Aerospace is one step closer to building an electric plane." https://heartaerospace.com/wp-content/uploads/2021/07/Heart-Aerospace-Series-A-Press-Release-July-13-2021.pdf.

Heritage Foundation. (2022). "Elaine Chao, Former Distinguished Fellow." https://www.heritage.org/staff/elaine-chao.

Hoffman, Adonis. (2021). "President Biden, oil and gas power America. Now is not the time to bite the hand of Big Oil." Fox News. https://www.foxnews.com/opinion/biden-oil-and-gas-power-america-big-oil-adonis-hoffman.

Holder, Sarah. (2019). "Pete Buttigieg's Climate Vision: Local Fixes for a Planet in Crisis." *Bloomberg.* https://www.bloomberg.com/news/articles/2019-11-04/inside-pete-buttigieg-s-2-trillion-climate-plan.

Hyatt, Kyle. (2018). "A Republican senator is trying to kill the federal EV tax credit, again." *CNET Road/Show*. https://www.cnet.com/roadshow/news/republican-kill-ev-tax-credit-bill/.

Illinois.gov. (2021). "Gov. Pritzker Signs Transformative Legislation Establishing Illinois as a National Leader on Climate Action." https://www.illinois.gov/news/press-release.23893.html.

Inquirer staff. (2021). "Ida death toll rises in Philly region." *Philadelphia Inquirer*. https://www.inquirer.com/weather/live/ida-philadelphia-flooding-tornado-pennsylvania-new-jersey-20210902.html.

IRENA. (2020). "Renewables increasingly beat even cheapest coal competitors on cost." https://www.irena.org/newsroom/pressreleases/2020/Jun/Renewables-Increasingly-Beat-Even-Cheapest-Coal-Competitors-on-Cost.

Irwin, Douglas A. (2004). "The Aftermath of Hamilton's 'Report on Manufactures.'" *Journal of Economic History,* Vol. 64, No., 3, 800–821. https://www.jstor.org/stable/3874820.

James, LeBron (@kingjames). (2019). "Man, these LA fires aren't no joke." Twitter. https://twitter.com/kingjames/status/1188770703438278657.

Jefferson, Thomas. (1972). "From Thomas Jefferson to George Washington." National Archives. https://founders.archives.gov/documents/Jefferson/01-24-02-0330.

Jessop, Simon, Katy Daigle, and Valerie Volcovici. (2012). "Baffled? You're not alone—COP26 spawns confusing array of acronyms." *Reuters*. https://www.reuters.com/business/cop/mission-coalition-cop26-spawns-confusing-clusters-2021-11-05/.

Johnson Hess, Abigail. (2017). "How Tesla and Elon Musk became household names." *CNBC*. https://www.cnbc.com/2017/11/21/how-tesla-and-elon-musk-became-household-names.html.

Joiner West, Erica. (2021). "Hitachi Automotive to make electric motors in Kentucky." Southern Automotive Alliance. https://southernautomotivealliance.com/hitachi-automotive-to-make-electric-motors-in-kentucky/.

Jordan, Jim (@Jim_Jordan). (2021). "Joe Biden is the new Jimmy Carter." Twitter. https://twitter.com/jim_jordan/status/1392099930978803717.

Kane, Paul. (2019). "A beloved senator heads into retirement with no clear successor to fill the void." *Washington Post*. https://www.washingtonpost.com/powerpost/a-beloved-senator-heads-into-retirement-with-no-clear-successor-to-fill-the-void/2019/12/05/6922d9b6-16b4-11ea-9110-3b34ce1d92b1_story.html.

Kaplan, Sarah, and Andrew Ba Tran. (2021). "Nearly 1 in 3 Americans experienced a weather disaster this summer." *Washington Post*. https://www.washingtonpost.com/climate-environment/2021/09/04/climate-disaster-hurricane-ida/.

Kaufman, Leslie. (2021). "Millions in Fire-Ravaged California Risk Losing Home Insurance." *Bloomberg*. https://www.bloomberg.com/news/articles/2021-09-25/millions-in-fire-ravaged-california-risk-losing-home-insurance.

Kelley, Alexandra. (2021). "Damage from Ida estimated to cost $18 billion." *The Hill*. https://thehill.com/changing-america/resilience/natural-disasters/570493-damage-from-ida-estimated-to-cost-18-billion.

Kelley, Colin, et al. (2015). "Climate Change in the Fertile Crescent and Implications of the Recent Syrian Drought." Proceedings of the National Academy of Sciences. https://www.pnas.org/content/112/11/3241.

Kelley Blue Book. (2021). "Average New-Vehicle Transaction Prices Top $54,000 For First Time." https://mediaroom.kbb.com/2021-10-13-Average-New-Vehicle-Transaction-Prices-Top-45,000-for-First-Time,-According-to-Kelley-Blue-Book.

Kelley Blue Book. (2021). "Electrified Vehicles Sales Boom in Q3." https://mediaroom.kbb.com/2021-10-26-Electrified-Vehicle-Sales-Boom-in-Q3-to-Surge-Past-One-Million-Units-in-2021,-According-to-Kelley-Blue-Book-Report.

Kennedy, Brian, and Alison Spencer. (2021). "Most Americans support expanding solar and wind energy, but Republican support has dropped." Pew Research Center. https://www.pewresearch.org/fact-tank/2021/06/08/most-americans-support-expanding-solar-and-wind-energy-but-republican-support-has-dropped/.

Kennedy, John F. (1961). "Address to Joint Session of Congress May 25, 1961." John F. Kennedy Presidential Library and Museum. https://www.jfklibrary.org/learn/about-jfk/historic-speeches/address-to-joint-session-of-congress-may-25-1961.

Khan, Shariq, and Scott DiSavino. (2021). "NextEra doubles down on renewables as energy transition gathers steam." *Reuters*. https://www.reuters.com/article/us-nextera-energy-results/nextera-doubles-down-on-renewables-as-energy-transition-gathers-steam-idUSKBN29V1LM.

Kingsley, Patrick. (2018). "Trump Says California Can Learn from Finland on Fires. Is He Right?" *New York Times*. https://www.nytimes.com/2018/11/18/world/europe/finland-california-wildfires-trump-raking.html.

Kohler, Judith. (2019). "Gov. Polis signs bills on renewable energy, but what does that mean for Colorado's energy future?" *Denver Post*. https://www.denverpost.com/2019/05/30/colorado-jared-polis-renewable-energy-climate/.

Koran, Mario. "California power outages could cost region more than $2bn, some experts say." *The Guardian*. https://www.theguardian.com/us-news/2019/oct/11 california-power-outages-cost-business-wildfires.

Koren, James Rufus. (2018). "Insurer Merced went belly up after Camp fire. Here's what policyholders need to know." *Los Angeles Times*. https://www.latimes.com/business/la-fi-merced-insurance-customers-20181204-story.html.

Krauss, Clifford, and Peter Eavis. (2021). "Climate Activists Defeat Exxon in Push for Clean Energy." *New York Times*. https://www.nytimes.com/2021/05/26/business/exxon-mobil-climate-change.html.

Kusnetz, Nicholas. (2017). "Rising Seas Are Flooding Norfolk Naval Base, and There's No Plan to Fix It." *Inside Climate News*. https://insideclimatenews.org/news/25102017/military-norfolk-naval-base-flooding-climate-change-sea-level-global-warming-virginia/.

Lavelle, Marianne. (2017). "Rick Perry Denies Climate Change Role of CO2." *Inside Climate News*. https://insideclimatenews.org/news/19062017/rick-perry-dismisses-co2-climate-change-oceans-environment-cnbc-kernen/.

Lazard. (2019). "Levelized Cost of Energy and Levelized Cost of Storage, 2019." https://www.lazard.com/perspective/lcoe2019.

Lazard. (2021). "Levelized Cost of Energy, Levelized Cost of Storage, and Levelized Cost of Hydrogen." https://www.lazard.com/perspective/levelized-cost-of-energy-levelized-cost-of-storage-and-levelized-cost-of-hydrogen/.

Lemire, Jonathan, et al. (2020). "Trump spurns science on climate: 'Don't think science knows.'" *Associated Press*. https://apnews.com/article/climate-climate-change-elections-joe-biden-campaigns-bd152cd786b58e45c61bebf2457f9930.

Liewer, Steve. (2019). "Flood Recovery at Offutt could cost $1 billion and take five years." *Omaha World Herald*. https://omaha.com/local/flood-recovery-at-offutt-could-cost-1-billion-and-take-five-years/article_8f4fff1a-ff4e-5265-bf73-3d6df6be94e8.html.

Linton, Caroline. (2020). "Lightning Siege hits California with nearly 12,000 strikes in a week," *CBS News*. https://www.cbsnews.com/news/lightning-siege-hits-california-with-nearly-12000-strikes-in-a-week-2020-08-22/.

Lipton, Eric. (2020). "'The Coal Industry Is Back,' Trump Proclaimed. It Wasn't." *New York Times*. https://www.nytimes.com/2020/10/05/us/politics/trump-coal-industry.html.

Little, Jacob. (2015). "Microsoft's Carbon Fee Model Presented by Alumna Tamara DiCaprio." *Daily Nexus*. https://dailynexus.com/2015-10-16/microsofts-carbon-fee-model-presented-by-alumna-tamara-dicaprio/.

Lowerccarbon Capital. (2022). "Kickass Companies." https://lowercarboncapital.com/companies/.

Lunsford, Mackensy. (2018). "Florence will cost farmers more than $1 billion in lost crops and livestock." *Asheville Citizen-Times*. https://www.citizen-times.com/story/news/local/2018/09/26/agricultural-losses-become-clearer-eastern-north-carolina-recovers-hurricane-florence/1410031002/.

Makower, Joel. (2021). "The business of COP is business." *GreenBiz*. https://www.greenbiz.com/article/business-cop-business.

Malloy, Daniel. (2015). "Johnny Isakson tussles with a Democrat on Senate Floor: 'I don't believe climate change is a religion.'" *Atlanta Journal-Constitution*. https://www.ajc.com/blog/politics/johnny-isakson-tussles-with-democrat-senate-floor-don-believe-climate-change-religion/ajruIc3fbNCxmIHFNplS8J/.

Martin, Jonathan, and Alexander Burns. (2021). "Reeling From Surprise Losses, Democrats Sound the Alarm for 2022." *New York Times*. https://www.nytimes.com/2021/11/03/us/politics/democrat-losses-2022.html.

McCammond, Alexi. (2021). "AOC, The Squad defend infrastructure 'no' vote." *Axios*. https://www.axios.com/aoc-squad-defend-infrastructure-no-vote-69729c85-10fe-4e71-b0c3-8f99d936626f.html.

McConnell, Mitch. (2021). "'Piecemeal Green New Deal' Putting Ideology Ahead of Working Americans." https://www.republicanleader.senate.gov/newsroom/remarks/piecemeal-green-new-deal-putting-ideology-ahead-of-working-americans.

McKinsey & Co. (2017). "An integrated perspective on the future of mobility, part 2: transforming urban delivery." https://www.mckinsey.com/~/media/mckinsey/business%20functions/sustainability/our%20insights/urban%20commercial%20transport%20and%20the%20future%20of%20mobility/an-integrated-perspective-on-the-future-of-mobility.pdf.

McKinsey & Co. (2020). "Charging electric-vehicle fleets: How to seize the emerging opportunity." https://www.mckinsey.com/~/media/McKinsey/Business%20Functions/Sustainability/Our%20Insights/Charging%20electric%20vehicle%20fleets%20How%20to%20seize%20the%20emerging%20opportunity/Charging-electric-vehicle-fleets-how-to-seize-the-emerging-opportunity-FINAL.pdf.

Milieudefensie et al. v. Royal Dutch Shell judgement. (2021). https://uitspraken.rechtspraak.nl/inziendocument?id=ECLI:NL:RBDHA:2021:5339.

Military.com. (2022). "Naval Station Norfolk Base Guide." https://www.military.com/base-guide/naval-station-norfolk.

Moleski, Vincent. (2020). "First Ever Fire Tornado Warning Issued in CA." *The Sacramento Bee*. https://www.firehouse.com/operations-training/wildland/news/21150367/first-ever-fire-tornado-warning-issued-in-ca.

Moody's Investors Service. (2021). "Research Announcement: Moody's—Financial firms that take rapid, predictable pace to zero financed emissions will win the race." https://www.moodys.com/research/Moodys-Financial-firms-that-take-rapid-predictable-pace-to-zero--PBC_1305598.

Morehouse, Catherine. (2019). "Three's company: New Mexico joins California, Hawaii in approving 100% clean energy mandate." *Utility Dive*. https://www.utilitydive.com/news/threes-company-new-mexico-joins-california-hawaii-in-approving-100-clea/550390/.

Morse, Ian. (2020). "Down on the Farm That Harvests Metal from Plants." *New York Times*. https://www.nytimes.com/2020/02/26/science/metal-plants-farm.html.

Mottley, Mia. (2021). "Speech, Mia Mottley, Prime Minister of Barbados at the Opening of the Cop26 World Leaders Summit." United Nations Climate Change. https://www.youtube.com/watch?v=PN6THYZ4ngM.

Mufson, Steven. (2018). "Scott Pruitt's likely successor has long lobbying history on issues before the EPA." *Washington Post*. https://www.washingtonpost.com/business/economy/epas-acting-administrator-has-long-lobbying-record-on-issues-before-the-agency/2018/07/05/a591cd40-6a6b-11e8-bea7-c8eb28bc52b1_story.html.

Mufson, Steven. (2019). "Trump's pick for managing federal lands doesn't believe the government should have any." *Washington Post*. https://www.washingtonpost.com/climate-environment/trumps-pick-for-managing-federal-lands-doesnt-believe-the-government-should-have-any/2019/07/31/0bc1118c-b2cf-11e9-8949-5f36ff92706e_story.html.

Mufson, Steven. (2021). "What could finally stop new coal plants? Pulling the plug on their insurance." *Washington Post*. https://www.washingtonpost.com/climate-environment/2021/10/26/climate-change-insurance-coal/.

Mufson, Steven. (2021). "Why has Andy Karsner frightened the mighty Exxon-Mobil?" *Washington Post*. https://www.washingtonpost.com/climate-environment/2021/06/19/exxon-board-karsner-engine1/.

Munich RE. (2021). "Record hurricane season and major wildfires—the natural disaster figures for 2020." https://www.munichre.com/en/company/media-relations/media-information-and-corporate-news/media-information/2021/2020-natural-disasters-balance.html.

NASA. (2019). "Apollo 11, July 20, 1969, One Giant Leap for Mankind." https://www.nasa.gov/mission_pages/apollo/apollo11.html.

NASA. (2022). "Fact Sheets NASA Langley Research Center's Contribution to the Apollo Program." https://www.nasa.gov/centers/langley/news/factsheets/Apollo.html.

National Interagency Fire Center. (2020). "2020 National Large Incident Year-to-Date Report." https://web.archive.org/web/20201229021815/https://gacc.nifc.gov/sacc/predictive/intelligence/NationalLargeIncidentYTDReport.pdf.

National Interagency Fire Center. (2021). "2021 National Large Incident Year-to-Date Report." https://gacc.nifc.gov/sacc/predictive/intelligence/NationalLargeIncidentYTDReport.pdf.

National Snow & Ice Data Center. (2021). "Arctic sea ice has reached minimum extent for 2021." https://nsidc.org/news/newsroom/arctic-sea-ice-has-reached-minimum-extent-2021.

NC Clean Energy Technology Center. (2020). "DSIRE Renewable & Clean Energy Standards." https://ncsolarcen-prod.s3.amazonaws.com/wp-content/uploads/2020/09/RPS-CES-Sept2020.pdf.

NC Governor Roy Cooper. (2018). "Updated Estimates Show Florence Caused $17 Billion in Damage." https://governor.nc.gov/news/updated-estimates-show-florence-caused-17-billion-damage.

NC Governor Roy Cooper. (2021). "Governor Cooper Signs Energy Bill Including Carbon Reduction Goals into Law." https://governor.nc.gov/news/press-releases/2021/10/13/governor-cooper-signs-energy-bill-including-carbon-reduction-goals-law.

NextEra Energy. (2021). "Did You Know?" https://www.nexteraenergy.com/company/subsidiaries.html.

NextEra. https://www.nexteraenergy.com/.

Nextera.com. (2022). https://www.nexteraenergyresources.com/what-we-do/wind.html.

NIFC.gov. (2022). https://www.nifc.gov/.

Nilsen, Ella. (2020). "Joe Biden and Bernie Sanders are building new, policy-focused task forces." *Vox*. https://www.vox.com/2020/5/13/21257078/joe-biden-bernie-sanders-joint-unity-task-forces-democratic-policy.

NOAA, Office of Response and Restoration. (2016). "1976: A Winter of Ship Accidents." https://response.restoration.noaa.gov/oil-and-chemical-spills/significant-incidents/argo-merchant-oil-spill/1976-winter-ship-accidents.htm.

NOAA. (2020). "2020 Atlantic Hurricane Season takes infamous top spot for busiest on record." https://www.noaa.gov/news/2020-atlantic-hurricane-season-takes-infamous-top-spot-for-busiest-on-record.

NOAA. (2022). "Billion-Dollar Weather and Climate Disasters." https://www.ncdc.noaa.gov/billions/.

NOAA et al. (2018). "Fourth National Climate Assessment." https://nca2018.globalchange.gov/.

NRDC. (2013). "California's Energy Efficiency Success Story." https://www.nrdc.org/sites/default/files/ca-success-story-FS.pdf.

O'Connor, Amy. (2020). "Florida Property Insurance Market Inches Closer to Crisis." *Insurance Journal*. https://www.insurancejournal.com/news/southeast/2020/10/29/588564.htm.

O'Dell, John. (2001). "GM Sues to Overturn State's Zero Emission Vehicle Mandate." *Los Angeles Times*. https://www.latimes.com/archives/la-xpm-2001-feb-24-fi-29699-story.html.

Obama White House. (2011). "Remarks by the President in State of the Union Address." https://obamawhitehouse.archives.gov/the-press-office/2011/01/25/remarks-president-state-union-address.

Office of Governor Gavin Newsom. (2021). "Ahead of Peak Fire Season, Governor Newsom Announces Surge in Firefighting Support." https://www.gov.ca.gov/2021/03/30/ahead-of-peak-fire-season-governor-newsom-announces-surge-in-firefighting-support-3-30-21/.

OpenSecrets. (2022). "Sen. John A. Barrasso – Campaign Finance Summary." https://www.opensecrets.org/members-of-congress/summary?cid=N00006236&cycle=CAREER.

Oregonian staff. (2020). "Oregon Wildfires Destroyed More Than 4,000 Homes. Here's Where." *Oregonian*. https://www.oregonlive.com/wildfires/2020/10/oregon-wildfires-destroyed-more-than-4000-homes-heres-where.html.

Pavley, Fran. (2021). Interview.

Peltz, James F. (2019). "PG&E to file for bankruptcy as wildfire costs hit $30 billion." *Los Angeles Times*. https://www.latimes.com/california/story/2019-12-08/wildfire-firefighter-overtime-budget.

Pollack, Nicole. (2021). "Six new wind farms proposed for Wyoming." *Casper Star-Tribune*. https://trib.com/business/energy/six-new-wind-farms-proposed-for-wyoming/article_41cbf2db-e482-5bb0-a629-fbabdb7296da.html.

Ponton, Brendan. (2020). "City of Hampton uses innovative tool to fight flooding." 3*WTKR*. https://www.wtkr.com/news/city-of-hampton-uses-innovative-tool-to-fight-flooding.

Popovich, Nadja, et al. (2021). "The Trump Administration Rolled Back More Than 100 Environmental Rules. Here's the Full List." *New York Times*. https://www.nytimes.com/interactive/2020/climate/trump-environment-rollbacks-list.html.

PPCA. (2021). "New PPCA members tip the scales toward 'consigning coal to history.' COP26." https://www.poweringpastcoal.org/news/press-release/new-ppca-members-tip-the-scales-towards-consigning-coal-to-history-at-cop26.

Preston, Benjamin. (2020). "Consumer Reports Survey Shows Strong Interest in Electric Cars." *Consumer Reports*. https://www.consumerreports.org/hybrids-evs/cr-survey-shows-strong-interest-in-evs-a1481807376/.

Quinson, Tim. (2021). "Wall Street's $22 Trillion Carbon Time Bomb." *Financial Post*. https://financialpost.com/pmn/business-pmn/wall-streets-22-trillion-carbon-time-bomb.

Resources for the Future. (2021). "Wildfires in the United States 101: Context and Consequences." https://www.rff.org/publications/explainers/wildfires-in-the-united-states-101-context-and-consequences/.

Reuters. (2021). "Volvo Group says launches world's first fossil-free steel vehicle." https://www.reuters.com/business/sustainable-business/volvo-trucks-says-launches-worlds-first-fossil-free-steel-vehicle-2021-10-13/.

Rhodes, Shawn C. (2004). "Easy Company Carries on Tradition of Excellence." U.S. Marines. https://www.lejeune.marines.mil/News/Article/Article/511982/easy-company-carries-on-tradition-of-excellence/.

Rhodium Group. (2021). "Pathways to Build Back Better: Nearly a Gigaton on the Table in Congress." https://rhg.com/research/build-back-better-congress-budget/.

Rice, Doyle. (2019). "USA had world's 3 costliest natural disasters in 2018, and Camp Fire was the worst." *USA Today*. https://www.usatoday.com/story/news/2019/01/08/natural-disasters-camp-fire-worlds-costliest-catastrophe-2018/2504865002/.

Roberts, David. (2016). "Florida's outrageously deceptive solar ballot initiative, explained." *Vox*. https://www.vox.com/science-and-health/2016/11/4/13485164/florida-amendment-1-explained.

Rockefeller Foundation. (2021). "Historic Alliance Launches at COP26 to Accelerate a Transition to Renewable Energy, Access to Energy for All, and Jobs." https://www.rockefellerfoundation.org/news/historic-alliance-launches-at-cop26-to-accelerate-renewable-energy-climate-solutions-and-jobs/.

Roosevelt, Margot. (2010). "How Fran Pavley changed the American auto industry." *Los Angeles Times*. https://latimesblogs.latimes.com/greenspace/2010/04/fuel-economy-global-warming.html.

Roosevelt, Theodore. (1908). "Conservation as a National Duty." *Voices of Democracy*. http://voicesofdemocracy.umd.edu/theodore-roosevelt-conservation-as-a-national-duty-speech-text/.

Root, Al. (2021). "Ford's Electric F-150 Hits 130,000 Reservations. How That Stacks Up against Other EV Makers." *Barron's*. https://www.barrons.com/articles/ford-electric-f-150-reservations-tesla-ev-makers-51630674776.

Russell, Clyde. (2021). "Coal trajectory is set whether it's "'phase out' or 'phase down.'" *Reuters*. https://www.reuters.com/business/cop/coal-trajectory-is-set-whether-its-phase-out-or-phase-down-russell-2021-11-14/.

Rust, Susanne, and Tony Barboza. (2020). "How climate change is fueling record-breaking California wildfires, heat and smog." *Los Angeles Times*. https://www.latimes.com/california/story/2020-09-13/climate-change-wildfires-california-west-coast.

S&P Global. (2020). "Investment in US clean energy to total $55 billion in 2020." https://www.spglobal.com/platts/en/market-insights/latest-news/electric-power/112320-investment-in-us-clean-energy-to-total-55-bil-in-2020-generate-capital.

Sacca, Chris. (2021). "Cash Cools Everything Around Me." Lowercarbon Capital. https://lowercarboncapital.com/2021/08/12/cash-cools-everything-around-me/.

Salam, Erum. (2021). "What's actually in Biden's Build Back Better bill? And how would it affect you?" *The Guardian*. https://www.theguardian.com/us-news/2021/oct/18/what-is-build-back-better-crash-course.

Samenow, Jason. (2020). "Cedar Rapids and nearby Iowa communities, still in shambles days after destructive derecho, plead for help." *Washington Post*. https://www.washingtonpost.com/weather/2020/08/14/cedar-rapids-iowa-derecho/.

Schatzker, Erik. (2021). "Hertz Order for 100,000 EVs Sends Tesla Value to $1 Trillion." *Bloomberg*. https://www.bloomberg.com/news/articles/2021-10-25/hertz-said-to-order-100-000-teslas-in-car-rental-market-shake-up.

Schogol, Jeff. (2016). "Marine recruits being evacuated from Parris Island ahead of Hurricane Matthew." *Marine Corps Times*. https://www.marinecorpstimes.com/news/your-marine-corps/2016/10/05/marine-recruits-being-evacuated-from-parris-island-ahead-of-hurricane-matthew/.

Schwartz, Matthew. (2020). "Iowa Derecho This August Was Most Costly Thunderstorm Event in Modern U.S. History." *NPR*. https://www.npr.org/2020/10/18/925154035/iowa-derecho-this-august-was-most-costly-thunderstorm-event-in-modern-u-s-histor#:~:text=More%20Podcasts%20%26%20Shows-,This%20Summer's%20Iowa%20Thunderstorms%20Were%20Some%20Of%20The%20Costliest%20On,more%20expensive%20than%20some%20hurricanes.

Science Based Targets. (2022). https://sciencebasedtargets.org/.

Searching In History. (2015). "Samuel Slater: The Father of American Industrial Revolution." https://searchinginhistory.blogspot.com/2015/01/samuel-slater-father-of-american.html.

SEIA.org. (2022). "North Carolina Solar." https://www.seia.org/states-map.

Senate Armed Services Committee. (2018). Hearing, "Navy and Marine Corps Readiness." https://www.c-span.org/video/?455659-1/marine-corps-commandant-navy-secretary-testify-readiness.

Siler, Wes. (2019). "The David Bernhardt Scandal Tracker." *Outside*. https://www.outsideonline.com/outdoor-adventure/environment/david-bernhardt-scandal-tracker/.

Silva, Rick. (2018). "The Humboldt Fire, 10 years later." *Enterprise-Record*. https://www.chicoer.com/2018/06/15/humboldt-fire-10-years-later/.

Smith, Allan. (2021). "McConnell says he's '100 percent' focused on 'stopping' Biden's administration." *NBC News*. https://www.nbcnews.com/politics/joe-biden/mcconnell-says-he-s-100-percent-focused-stopping-biden-s-n1266443.

Smith, Brad. (2020). "Microsoft will be carbon negative by 2030." Official Microsoft Blog. https://blogs.microsoft.com/blog/2020/01/16/microsoft-will-be-carbon-negative-by-2030/.

Snow, Shawn. (2018). "$3.6 billion price tag to rebuild Lejeune buildings damaged by Hurricane Florence." *Marine Times*. https://www.marinecorpstimes.com/news/your-marine-corps/2018/12/12/36-billion-price-tag-to-rebuild-lejeune-buildings-damaged-by-hurricane-florence/.

Snow, Shawn. (2018). "Parris Island to evacuate ahead of Hurricane Florence." *Military Times*. https://www.militarytimes.com/news/your-marine-corps/2018/09/10/parris-island-to-evacuate-ahead-of-hurricane-florence/.

Southern Company. (2021). CEO Tom Fanning on Q32021 Results. https://seekingalpha.com/article/4465705-southern-company-ceo-tom-fanning-on-q3-2021-results-earnings-call-transcript.

Stanford University. (2020). "COVID lockdown causes record drop in carbon emissions for 2020." *Stanford Earth Matters*. https://earth.stanford.edu/news/covid-lockdown-causes-record-drop-carbon-emissions-2020#gs.h84938.

Stankiewicz, Alyssa. (2021). "The Number of New Sustainable Funds Hits an All-Time Record." *Morningstar*. https://www.morningstar.com/articles/1062299/the-number-of-new-sustainable-funds-hits-an-all-time-record.

Stayer, Philip A. (2018). "Hurricane Florence and the poultry industry: Coping with the aftermath." *Poultry Health Today*. https://poultryhealthtoday.com/mobile/article/?id=6315.

Stiles, Matt. (2019). "Firefighter overtime surged 65% in a decade, costing California $5 billion a year in wages." *Los Angeles Times*. https://www.latimes.com/california/story/2019-12-08/wildfire-firefighter-overtime-budget.

Surma, Katie. (2021). "Mary Nichols Was the Early Favorite to Run Biden's EPA, Before She Became a 'Casualty.'" *Inside Climate News*. https://insideclimatenews.org/news/09022021/mary-nichols-epa-joe-biden-michael-regan-environmental-justice/.

Swiss RE Group. (2019). "Secondary perils to wreak evermore natural catastrophe devastation globally." https://www.swissre.com/media/news-releases/nr-20190410-sigma-2-2019.html.

Swiss RE Group. (2021). "The Economics of Climate Change." https://www.swissre.com/institute/research/topics-and-risk-dialogues/climate-and-natural-catastrophe-risk/expertise-publication-economics-of-climate-change.html.

Telle, Svenja. (2021). "COP26 and the climate finance bubble." *PitchBook*. https://pitchbook.com/news/articles/cop26-2021-climate-change-finance-bubble.

Tesla. (2013). "Tesla Repays Department of Energy Loan Nine Years Early." https://www.tesla.com/blog/tesla-repays-department-energy-loan-nine-years-early.

Texas Public Policy Foundation. (2015). "Doug Domenech joins TPPF as Director of Fueling Freedom Project." https://www.texaspolicy.com/press/doug-domenech-joins-tppf-as-director-of-fueling-freedom-project.

The Eagle. (2019). "Weather Whys: The Dust Bowl." https://theeagle.com/townnews/agriculture/weather-whys-the-dust-bowl/article_abdd3290-8c11-11e2-a59a-0019bb2963f4.html.

The White House. (2021). "Fact Sheet: The Bipartisan Infrastructure Deal." https://www.whitehouse.gov/briefing-room/statements-releases/2021/11/06/fact-sheet-the-bipartisan-infrastructure-deal/.

The White House. (2021). "ICYMI: More than 1,000 Business Leaders Call on Congress to Build Back Better on Climate." https://e2.org/wp-content/uploads/2021/10/White-House-More-Than-1000-Business-Leaders-Call-on-Congress-to-Build-Back-Better-on-Climate.pdf.

The White House. (2021). "President Biden Delivers Remarks on the Passage of the Bipartisan Infrastructure Bill." https://www.youtube.com/watch?v=baFaMfPjqT0.

The White House. (2021). "Remarks by President Biden in Address to a Joint Session of Congress." https://www.whitehouse.gov/briefing-room/speeches-remarks/2021/04/29/remarks-by-president-biden-in-address-to-a-joint-session-of-congress/.

Tillis, Thom. (2020). "Tillis, Colleagues Urge Senate Leaders to Consider Clean Energy Jobs Impacted by Pandemic in Next COVID-19 Package." https://www.tillis.senate.gov/2020/7/tillis-colleagues-urge-senate-leaders-to-consider-clean-energy-jobs-impacted-by-pandemic-in-next-covid-19-package.

Tomlinson, Victoria. (2020). "UPS invests in Arrival and orders 10,000 Generation 2 Electric Vehicles." *Arrival.com*. https://arrival.com/us/en/news/ups-invests-in-arrival-and-orders-10000-generation-2-electric-vehicles.

Trevithick, Joseph. (2018). "A Tornado Left the USAF With Only One Active E-4B 'Doomsday Plane' for Months." *The Drive*. https://www.thedrive.com/the-war-zone/18996/a-tornado-left-the-usaf-with-only-one-active-e-4b-doomsday-plane-for-months.

Turner, Jay. (2014). "New EPA head Gina McCarthy proud of local roots." *Canton Citizen*. https://www.thecantoncitizen.com/2014/05/08/gina-mccarthy-profile/.

Turrentine, Jeff. (2017). "Conflicts of Interest Swamp the Department of Interior." *NRDC onEarth*. https://www.nrdc.org/onearth/conflicts-interest-swamp-department-interior.

Uhler, Andy. (2018). "Hurricane means fewer North Carolina sweet potatoes." *Marketplace*. https://www.marketplace.org/2018/11/22/nc-farmers-lick-their-wounds-after-hurricane-devastates-sweet-potato-crop/.

United Nations Climate Change Conference. (2022). "COP26 Goals." https://ukcop26.org/cop26-goals/.

Union of Concerned Scientists. (2016). "On the Front Lines of Rising Seas: Naval Station Norfolk, Virginia." https://www.ucsusa.org/resources/front-lines-rising-seas-naval-station-norfolk-virginia.

U.S. Bureau of Labor Statistics. (2021). "Solar and wind generation occupations: a look at the next decade." https://www.bls.gov/opub/btn/volume-10/solar-and-wind-generation-occupations-a-look-at-the-next-decade.htm.

U.S. Chamber of Commerce. (2021). "US Chamber Statement on Biden Climate Goal." https://www.uschamber.com/environment/us-chamber-statement-biden-climate-goal.

U.S. Chamber of Commerce. (2021). "US Chamber Vows to Defeat Reconciliation, Hails Voting Deadline for Bipartisan Infrastructure Deal." https://www.uschamber.com/infrastructure/us-chamber-vows-defeat-reconciliation-hails-voting-deadline-bipartisan-infrastructure.

USAID. (2022). "How the United States Benefits from Agricultural and Food Security Investments in Developing Countries." https://www.usaid.gov/sites/default/files/documents/1867/BIFAD_US_Benefit_Study.pdf.

USDA. (2020). Wildfire and Hurricane Indemnity Program-Plus (WHIP+). https://www.fsa.usda.gov/Assets/USDA-FSA-Public/usdafiles/FactSheets/2019/wildfire-and-hurricane-indemnity-program-plus_whip_august_2020.pdf.

USDA. (2022). "2020 State Agriculture Overview, North Carolina." https://www.nass.usda.gov/Quick_Stats/Ag_Overview/stateOverview.php?state=NORTH%20CAROLINA.

U.S. Department of Energy. (2021). "DOE Announced Up to $8.25 Billion in Loans to Enhance Electrical Transmission Nationwide." https://www.energy.gov/articles/doe-announces-825-billion-loans-enhance-electrical-transmission-nationwide.

U.S. Department of Energy. (2021). "Electric Vehicle Registrations by State." https://afdc.energy.gov/data/10962.

U.S. Department of Energy. (2021). "Land-Based Wind Market Report: 2021 Edition." https://www.energy.gov/sites/default/files/2021-08/Land-Based%20Wind%20Market%20Report%202021%20Edition_Full%20Report_FINAL.pdf.

U.S. Department of Energy. (2021). "Remarks as delivered by Secretary Granholm at President Biden's Leaders Summit on Climate." https://www.energy.gov/articles/remarks-delivered-secretary-granholm-president-bidens-leaders-summit-climate.

U.S. Department of Energy. (2022). "DOE Factsheet." https://www.energy.gov/articles/doe-fact-sheet-bipartisan-infrastructure-deal-will-deliver-american-workers-families-and-0.

U.S. Department of Energy. (2022). "Energy Star Impacts." https://www.energystar.gov/about/origins_mission/impacts.

U.S. Department of Energy. (2022). "Solar Energy in the United States." https://www.energy.gov/eere/solar/solar-energy-united-states.

U.S. Department of the Treasury. (2021). "Readout of Financial Stability Oversight Council Meeting on July 16, 2021." https://home.treasury.gov/news/press-releases/jy0278.

U.S. Department of the Treasury. (2021). "Remarks by Secretary of the Treasury Janet L. Yellen at the Venice International Conference on Climate." https://home.treasury.gov/news/press-releases/jy0271.

U.S. Department of the Treasury. (2021). "Remarks by Secretary of the Treasury Janet L. Yellen on the Executive Order on Climate-Related Financial Risks." https://home.treasury.gov/news/press-releases/jy0190.

U.S. Department of the Treasury. (2021). "Remarks by Secretary of the Treasury Janet L. Yellen at COP 26 in Glasgow, Scotland." https://home.treasury.gov/news/press-releases/jy0465.

U.S. Department of the Treasury. (2021). "Treasury Announces Coordinated Climate Policy Strategy with New Treasury Climate Hub and Climate Counselor." https://home.treasury.gov/news/press-releases/jy0134.

U.S. Department of Transportation. (2021). "US Department of Transportation announces $180 million funding opportunity for low or no emission transit vehicles & facilities." https://www.transit.dot.gov/about/news/us-department-transportation-announces-180-million-funding-opportunity-low-or-no.

U.S. Department of Transportation. (2022). "History of the Interstate Highway System." https://www.fhwa.dot.gov/interstate/history.cfm.

U.S. Energy Information Administration. (2021). "Iowa State Profile and Energy Estimates." https://www.eia.gov/state/analysis.php?sid=IA.

U.S. Environmental Protection Agency. (2021). "EPA Awards $10.5 Million to Clean Up 473 School Buses in 40 States." https://www.epa.gov/newsreleases/epa-awards-105-million-clean-473-school-buses-40-states.

U.S. Environmental Protection Agency. (2022). "Green Power Partnership 100% Green Power Users." https://www.epa.gov/greenpower/green-power-partnership-100-green-power-users.

U.S. Foreign Agricultural Service. (2022). "Soybeans." https://www.fas.usda.gov/commodities/soybeans.

U.S. Marines. (2022). "Commanding General Major General Julian D. Alford." https://www.trngcmd.marines.mil/Leaders/Leaders-View/Article/1875079/major-general-julian-d-alford/.

University of California, Irvine. (2020). "UCI, Tsinghua U.: California's 2018 wildfires caused $150 billion in damages." https://news.uci.edu/2020/12/07/uci-tsinghua-u-californias-2018-wildfires-caused-150-billion-in-damages/.

Van der Ent, Antony, et al. (2017). "Nickel biopathways in tropical nickel hyperaccumulating trees from Sabah (Malaysia)." *Nature.com*. https://www.nature.com/articles/srep41861.

Vanguard. (2021). "Vanguard Investment Stewardship Insights." https://about.vanguard.com/investment-stewardship/perspectives-and-commentary/Exxon_1663547_052021.pdf.

Virginia Coastal Policy Center. (2016). "Economic Toll of Sea Level Rise Could Be Significant in Hampton Roads, Study Finds." https://law.wm.edu/news/stories/2016/economic-toll-of-sea-level-rise-could-be-significant-in-hampton-roads-study-finds.php.

Wade, Will. (2021). "Tom Steyer Forms Fund for Global Fight Against Climate Change." *Bloomberg*. https://www.bloomberg.com/news/articles/2021-09-09/tom-steyer-forms-fund-for-global-fight-against-climate-change.

Wall Street Journal. (2021). "Kerry Addresses 'Phase Down' of Coal in COP26 Agreement." https://www.wsj.com/video/kerry-addresses-phase-down-of-coal-in-cop26-agreement/2803D978-DEDB-41B1-93CE-BD01D450A8D8.html.

Wall Street Journal, editorial board. (2021). "More Green Blackouts Ahead." https://www.wsj.com/articles/more-green-blackouts-ahead-11614125061.

Wayland, Michael. (2021). "General Motors plans to exclusively offer electric vehicles by 2035." *CNBC*. https://www.cnbc.com/2021/01/28/general-motors-plans-to-exclusively-offer-electric-vehicles-by-2035.html.

We Mean Business Coalition. (2021). "More Than 330 Major Businesses Call on US Congress to Build Back a More Resilient, Sustainable Economy from Covid-19." https://www.wemeanbusinesscoalition.org/press-release/more-than-330-major-businesses-call-on-u-s-congress-to-build-back-a-more-resilient-sustainable-economy-from-covid-19/.

Weisman, Jonathan, and Carl Hulse. (2021). "How a $1 Trillion Infrastructure Bill Survived an Intraparty Brawl." *New York Times*. https://www.nytimes.com/2021/11/06/us/politics/infrastructure-black-caucus-vote.html.

Welborn, Kyle. (2021). "Q3 2021 AgTech Venture Capital Investment and Exit Round Up." *CropLife News*. https://www.croplife.com/precision/q3-2021-agtech-venture-capital-investment-and-exit-round-up/.

Wells Fargo. (2021). "Wells Fargo surpasses $10 billion in tax-equity financing of renewable energy projects." https://stories.wf.com/wells-fargo-surpasses-10-billion-in-tax-equity-financing-of-renewable-energy-projects/.

Werner, Ben. (2018). "Camp Lejeune Marines May Shelter on Base During Hurricane Florence." *USNI News*. https://news.usni.org/2018/09/12/36499.

White House. (2021). "Leaders Summit on Climate Summary of Proceedings." https://www.whitehouse.gov/briefing-room/statements-releases/2021/04/23/leaders-summit-on-climate-summary-of-proceedings/.

White House. (2021). "Readout of the First National Climate Task Force Meeting." https://www.whitehouse.gov/briefing-room/statements-releases/2021/02/11/readout-of-the-first-national-climate-task-force-meeting/.

Wicker, Roger, and Dan Sullivan. (2020). "Polar icebreakers are key to America's national interest." *Defense News*. https://www.defensenews.com/opinion/commentary/2020/10/19/polar-icebreakers-are-key-to-americas-national-interest/.

Wigglesworth, Alex, and Joseph Serna. (2020). "California fire season shatters record with more than 4 million acres burned." *Los Angeles Times*. https://www.latimes.com/california/story/2020-10-04/california-fire-season-record-4-million-acres-burned.

Wirz, Matt. (2021). "British Hedge Fund Billionaire Takes Climate Fight to S&P 500." *Wall Street Journal*. https://www.wsj.com/articles/british-hedge-fund-billionaire-takes-climate-fight-to-s-p-500-11611842401.

Witzenburg, Gary. (2021). "GM's EV1 Electric Car Invented Many Technologies That Are Commonplace on Today's EVs." *Car and Driver*. https://www.caranddriver.com/news/a36887553/gm-ev1-electric-car-technology/.

World Economic Forum. (2021). "The Global Risks Report 2021." https://www.weforum.org/reports/the-global-risks-report-2021.

World Meteorological Organization and United Nations Office for Disaster Risk Reduction. (2021). "Climate and weather-related disasters surge five-fold over 50 years." https://news.un.org/en/story/2021/09/1098662.

World Resources Institute. (2020). "America's New Climate Economy: A Comprehensive Guide to the Economic Benefits of Climate Policy in the United States." https://www.wri.org/research/americas-new-climate-economy-comprehensive-guide-economic-benefits-climate-policy-united.

World Vision. (2018). "2018 Hurricane Florence: Facts, FAQs and How to Help." https://www.worldvision.org/disaster-relief-news-stories/2018-hurricane-florence-facts.

WTVD. (2015). "NC town rejects solar farm, fearing it would suck up all the energy from the sun." *WTVD ABC 11*. https://abc11.com/sun-solar-panels-energy/1122081/.

Yahoo Finance. (2020–2021). https://finance.yahoo.com/quote/fan/.

YouTube. (2021). "Gina McCarthy and Secretary Granholm Hit the Road in an EV." https://www.youtube.com/watch?v=g9pElxQpiQ8.

Zandi, Mark, and Bernard Yaros. (2021). "Macroeconomic Consequences of the Infrastructure Investment and Jobs Act & Build Back Better Framework." *Moody's Analytics*. https://www.moodysanalytics.com/-/media/article/2021/macro

economic-consequences-of-the-infrastructure-investment-and-jobs-act-and-build-back-better-framework.pdf.

ZeroAvia. (2021). "ZeroAvia Secures Additional $24.3 Million." https://www.prnewswire.com/news-releases/zeroavia-secures-additional-24-3-million-to-kick-off-large-engine-development-for-50-seat-zero-emission-aircraft-301259265.html.

Index

About the Author

Bob Keefe is executive director of E2, a national, nonpartisan group of business owners, investors, and professionals who leverage economic research and their business perspective to advance policies that are good for the environment and good for the economy. E2's national network includes more than eleven thousand business leaders spread across nine chapters, stretching from New York to Los Angeles, and a staff of advocates who work on climate and clean-energy policies at the federal and state levels. An affiliate of the Natural Resources Defense Council (NRDC), E2 is the foremost business voice on issues at the intersection of the environment and economy and the leading authority on clean-energy jobs in America.

Previously, Keefe spent nearly twenty-five years as a journalist. He covered business news news for the *St. Petersburg (Fla.) Times*; was technology editor for the *Austin (Tx) American-Statesman* and covered national technology news and the West for the *Cox Newspapers* chain. In 2008, he moved to Washington, D.C. to cover Congress and the White House for the *Atlanta Journal-Constitution*, before joining E2 in 2011. He resides in San Diego, California with his wife and daughters.